INTRODUCTORY
GEOLOGY

Introductory Geology

Caleb Price

Kruger Brentt
Publishers
2026

Kruger Brentt Publishers UK. LTD.
Company Number 9728962

Regd. Office: 68 St Margarets Road, Edgware, Middlesex HA8 9UU

ISBN: 978-1-78715-641-8

For information on all our publications visit our website at http://krugerbrentt.com/

PREFACE

Geology, the scientific study of the Earth, provides essential insights into the planet's structure, composition, history, and the dynamic processes that shape its surface. As societies increasingly face challenges related to natural hazards, resource management, and environmental change, a foundational understanding of geological principles becomes indispensable for students, researchers, and professionals across environmental and earth sciences.

This book, Introductory Geology, offers a clear and comprehensive introduction to the fundamental concepts that define the geological sciences. Designed for beginners and those seeking a structured entry point into the subject, it presents the core ideas, methods, and applications that form the basis for more advanced geological study.

The early chapters explore the nature of minerals and rocks, the building blocks of the Earth's crust, and the techniques used to classify and analyze them. Subsequent sections introduce key geological processes, including plate tectonics, volcanism, earthquakes, weathering, erosion, and sedimentation. The text also highlights the formation of landscapes, geological time scales, and the evolution of Earth's surface through millions of years.

Practical aspects of geology—such as field methods, map interpretation, and the study of fossils—are explained with clarity to help readers connect theoretical knowledge with real-world observations. Contemporary issues, including climate change, natural disaster assessment, and sustainable resource exploration, are addressed to demonstrate the relevance of geology in modern society.

By blending scientific explanation with accessible language, Introductory Geology serves as an essential foundation for anyone seeking to explore Earth's past, understand its present, and prepare for its future.

Caleb Price

CONTENTS

1 Chapter

PHYSICAL GEOLOGY: AN INTRODUCTION

1.1 INTRODUCTION: AN OVERVIEW

The typical basic geology student's understanding of the subject often centers on the memorizing of rocks and the examination of natural disasters; nevertheless, geology encompasses far more than these elements. Geology encompasses the examination of our globe, which is essential to our daily existence, influencing the energy we consume, the cultivation of our food, the structural integrity of our dwellings, and the resources utilized in the production of commonplace items such as metals and plastics. The optimal starting point for this course is to examine the methodologies employed to enhance our comprehension of the planet, the processes that mold it, and its historical context.

Science is not merely a collection of information to be memorized. Rather, it serves as a means to explore the surrounding world. You are probably already acquainted with the Scientific Method, but it is beneficial to revisit the procedure. The initial phase of the scientific method involves making an observation or acquiring background knowledge pertinent to the issue of interest. This can be accomplished by enrolling in courses on a certain subject, such as geology, or by meticulously observing and documenting your environment. Utilizing your knowledge and observations, you can build a hypothesis, which is a tested prediction regarding the functioning of a phenomenon. A hypothesis must be formulated to facilitate testing and falsification. Although it may seem peculiar, science endeavors to systematically refute a hypothesis, and only those that endure the scrutiny are deemed valid. This definition of a hypothesis is remarkable since the testing yields a novel observation that can subsequently be utilized to develop a new hypothesis. Consequently, regardless of whether the hypothesis is confirmed or refuted, it will yield fresh insights. The subsequent phase involves discussion with other scientists. This enables other scientists to replicate the experiment and modify it in innovative and previously unconsidered manners, thereby enhancing the original concept. These steps comprise the majority of the scientific method and the professional trajectory of any scientist. As numerous pertinent data and validated ideas amass, scientists can develop a theory. A scientific hypothesis is an explanation for a natural occurrence substantiated by extensive empirical evidence. A theory has not attained the status of a law due to ongoing discourse on its precise mechanisms or the underlying causes of a phenomenon, yet there is minimal contention regarding the existence of the subject in question.

This brings us back to Geology, the scientific examination of the Earth. Certain facets of geology are immediately testable, while others are not, necessitating that geologists employ creativity to uncover elements of the Earth and its history that remain beyond direct observation. This laboratory manual will address the constituents of the Earth (Minerals and Rocks), geological processes occurring both internally (Folds and Faults) and externally (Rivers and Climate), along with the theoretical framework that elucidates Earth's functionality (Plate Tectonics). An essential component of comprehending the Earth is an understanding of Geologic Time (the focus of the initial chapter), which aids in conceptualizing the rate and occurrence of geological events that have shaped the world as we recognize it today..

1.2 GEOLOGIC TIME

The duration of landscape carving, rock creation, and continental movement is a significant scientific inquiry. Various hypotheses on the Earth's age can fundamentally alter our understanding of the geological processes that shaped the planet. Short geologic time necessitates catastrophic events to create the surface features of the Earth, while an extensive duration permits the gradual processes seen in the present day.

Geologists have employed several methods to reconstruct geological time, aiming to chart significant events in Earth's history together with their durations. Researchers examining rocks successfully reconstructed a chronological sequence of geological formations to develop the Geologic Time Scale (Figure 1.1). This timescale was established by sequentially arranging rocks based on specific characteristics, like rock types, environmental indications, or fossils. Scientists examined evidence within the rocks and established their age comparatively. This procedure is termed Relative Dating, which involves ascertaining the comparative age of two items or events. You are younger than your parents. Your age or parental lineage is irrelevant, provided you can demonstrate that one individual is older than the other. Over time, scientists identified and refined methods to date specific rocks and the Earth itself. They determined that the Earth is around 4.54 billion years old and established a chronological framework for the geologic time scale. This procedure is termed Absolute Dating, which involves ascertaining the precise duration that has elapsed since an object was created or an event transpired.

Both absolute and relative dating has advantages and remain commonly employed by geologists. Utilizing relative dating to date rocks enables a geologist to efficiently recreate a sequence of events, frequently expeditiously and cost-effectively, applicable in fieldwork on rocky outcrops. Relative dating can be used to several rock types, but absolute dating is limited to certain minerals or materials. Nonetheless, absolute dating is the sole technique that enables scientists to assign a precise age to a specific rock.

Eon	Era	Period		Epoch	Boundary age
PHANEROZOIC	Cenozoic	Quaternary		Holocene	0 Yr. – 15K Yr
				Pleistocene	1.8 Ma
		Tertiary	Neogene	Pliocene	5.3 Ma
				Miocene	24 Ma
			Paleogene	Oligocene	34 Ma
				Eocene	55 Ma
				Paleocene	65 Ma
	Mesozoic	Cretaceous			142 Ma
		Jurassic			202 Ma
		Triassic			251 Ma
	Paleozoic	Permian			292 Ma
		Pennsylvanian	Carboniferous		320 Ma
		Mississippian			354 Ma
		Devonian			417 Ma
		Silurian			440 Ma
		Ordovician			495 Ma
		Cambrian			545 Ma
Proterozoic	Pre-Cambrian (general)				2500 Ma
Archean					4000 Ma
Hadean					4550 Ma

Figure 1.1 | The geologic time scale. Ma, Million years, K. Yr, Thousand years

1.2.1 Relative Time and Geologic Laws

Geologists employ methodologies grounded in Geologic Laws to determine relative time scales. A scientific law is a well-established and validated principle. In contrast to mathematics, demonstrating concepts in science is challenging; hence, Geologic Laws are typically straightforward and very uncomplicated. Prior to discussing the many geological rules, it is prudent to succinctly present the distinct types of rocks. Sedimentary rocks, such as sandstone, consist of fragmented remnants of other rocks that are eroded from elevated regions, transported by wind, ice, and water to lower elevations, and subsequently deposited. Igneous rocks are formed through the cooling and crystallization of molten rock. Ultimately, the application of heat and pressure to rocks results in the formation of metamorphic rocks.

This distinction is important because these three different rock types are formed differently and therefore, need to be interpreted differently.

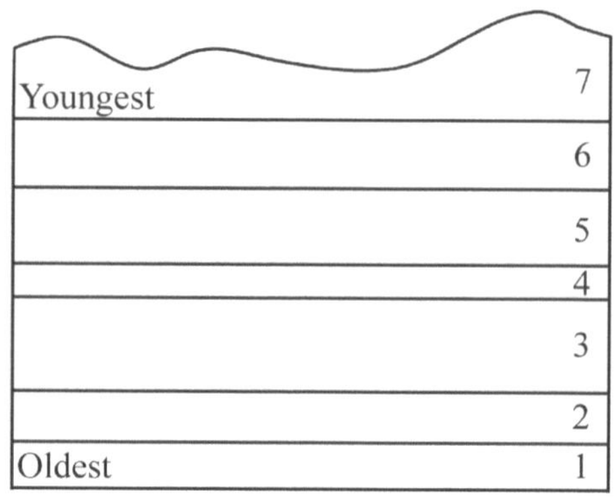

Figure 1.2 | Block diagram showing the relative age of sedimentary layers based on the Law of Superposition.

The Law of Superposition asserts that in an undisturbed sequence of sedimentary rocks, the oldest strata are located at the base, while the youngest strata are positioned at the top. Envision a river transporting sand into an ocean, where the sand will disperse along the ocean floor and settle atop the seabed. This sand was deposited subsequent to the deposition of the seafloor sand. A relative time scale of rock layers can be established, with the oldest rocks at the bottom (designated as #1 in Figure 1.2) and the youngest at the top of the outcrop (designated as #7 in Figure 1.2).

The Law of Original Horizontality asserts that undisturbed sedimentary rocks are deposited in a horizontal orientation. The deposition of sediment is governed by gravity, which exerts a downward force. Muddy water on a slope will flow downward and accumulate near the base rather than depositing along the hill. This indicates that sedimentary rock exhibiting tilting or folding was initially deposited in a horizontal orientation before undergoing subsequent deformation. (Figure 1.3).

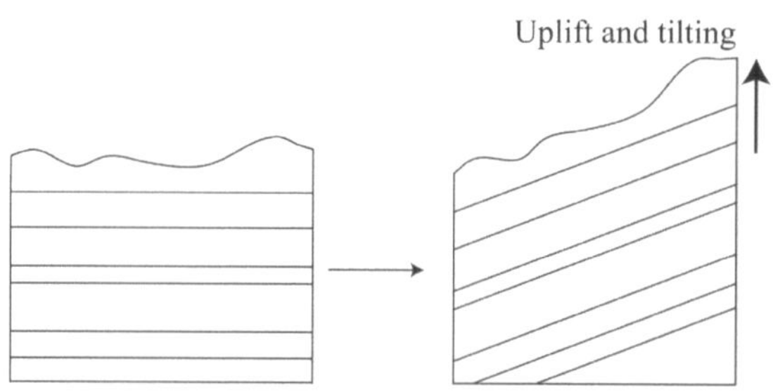

Figure 1.3 | Sedimentary rocks are deposited horizontally such that if the layers are tilted or folded it must have occurred following deposition.

The **Law of** Cross-**Cutting** states that when two geologic features intersect, the one that cuts across the other is younger. In essence, a feature has to be present before something can affect it. For example, if a fault fractures through a series of sedimentary rocks those sedimentary rocks must be older than the fault (Figure 1.4).

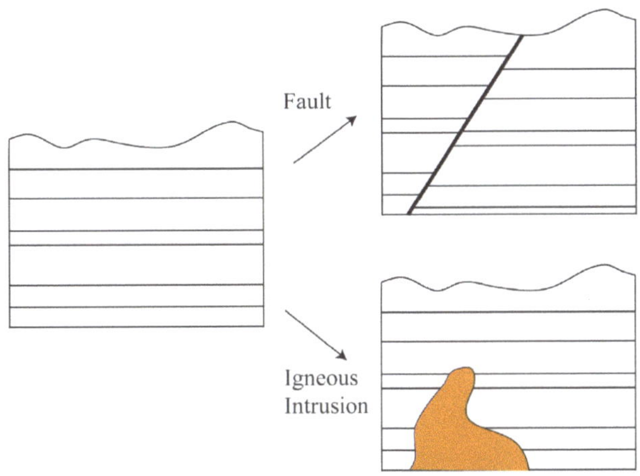

Figure 1.4 | Block diagrams showing the Law of Cross-Cutting. In both instances on the right the geological features (fault or Igneous intrusion) cut across the sedimentary layers and must then be younger.

Another significant aspect in constructing relative time scales is the absence of certain elements within a sequence of rocks. Unconformities are surfaces indicative of considerable weathering and erosion, leading to the absence or obliteration of geological time. Erosion frequently transpires in elevated regions such as continents or mountains, whereby the process of uplifting displaces rocks, leading to erosion and the degradation of a segment of a geological sequence; thus, significantly older rocks are revealed at the Earth's surface. If the area experiences subsidence, younger rocks will be deposited above the freshly exposed rocks. The duration of absence can be comparatively brief or may span billions of years. Three categories of unconformities exist, determined by the rocks situated above and below the unconformity (Figure 1.5). If the rock types differ above and below the unconformity, it is termed a Nonconformity. For instance, volcanic rock originating deep inside the Earth is raised and exposed at the surface, subsequently being overlain by sedimentary rock. If both the rocks above and below the erosion surface are sedimentary, the orientation of the strata is significant. An Angular Unconformity is defined as a surface when the rocks beneath are not parallel to those above. This frequently occurs due to the tilting or folding of underlying rocks before the erosion and deposition of the overlying younger rocks. If the strata above and below the erosion surface are parallel, the surface is termed a Disconformity.

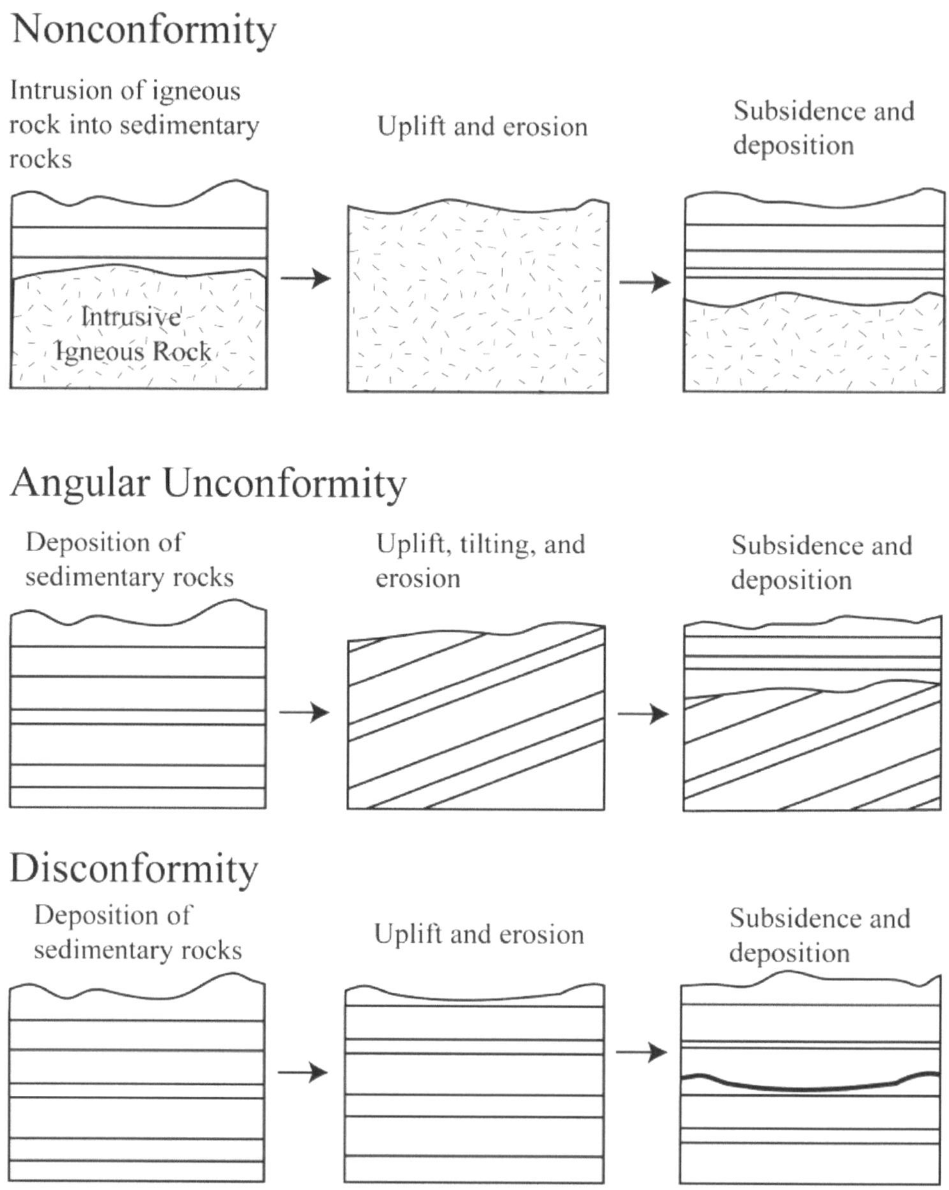

Figure 1.5 | Block diagrams showing the formation of the three types of Unconformities. The three unconformities differ based on the type of rock underneath the erosion surface.

Using these principles we can look at a series of rocks and determine their relative ages and even establish a series of events that must have occurred. Common events that are often recognized can include 1) Deposition of sedimentary layers, 2) Tilting or folding rocks, 3) Uplift and erosion of rocks, 4) Intrusion of liquid magma, and 5) Fracturing of rock (faulting). Figures 1.6 and 1.7 show how to piece together a series of geologic events using relative dating.

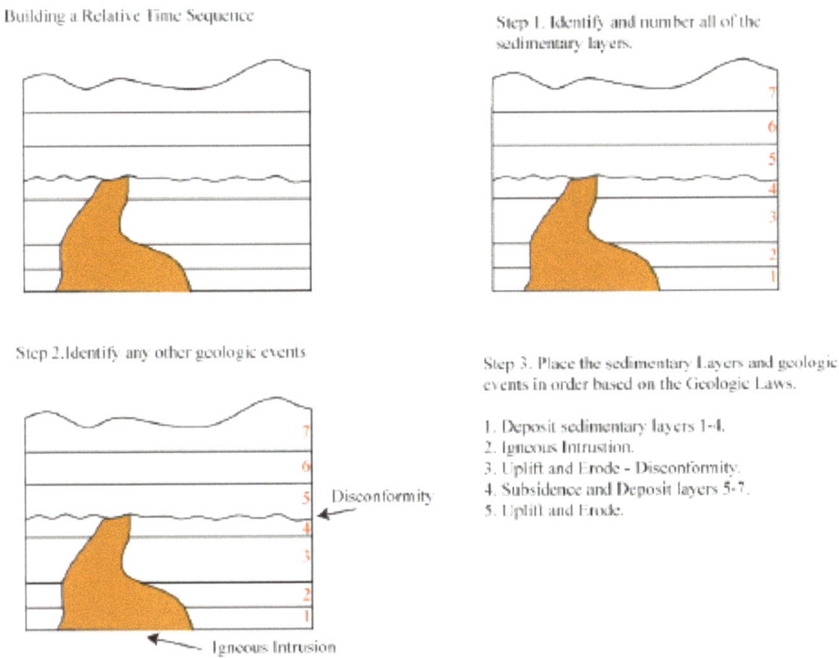

Figure 1.6 | An example showing how to determine a relative dating sequence of events from a block diagram.

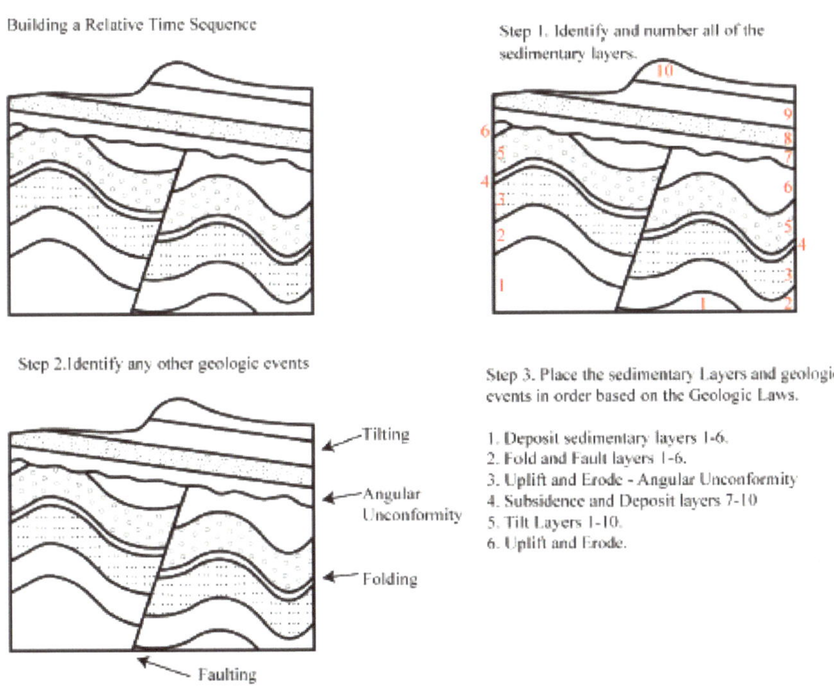

Figure 1.7 | An example showing how to determine a relative dating sequence of events from a block diagram.

1.2.2 LAB EXERCISE

Part A – Relative Time

Relative time is an important tool for geologist to quickly construct series of events, especially in the field. In the following section, apply what you have learned regarding relative time to the questions below.

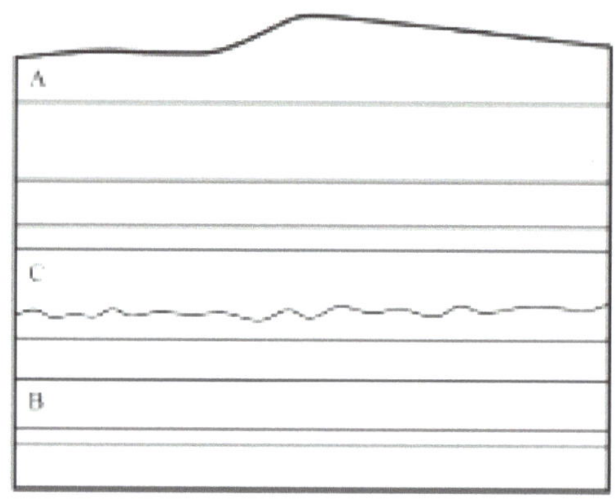

Figure 1.8 | Block diagram to use to answer questions 1 and 2.

1. In Figure 1.8, which of the following rock layers is the oldest?

   a. A b. B c. C

2. Which Geologic Law did you use to come to the conclusion you made in the previous question?

   a. The Law of Superposition

   b. The Law of Cross-Cutting

   c. The Law of Original Horizontality

   d. Unconformities

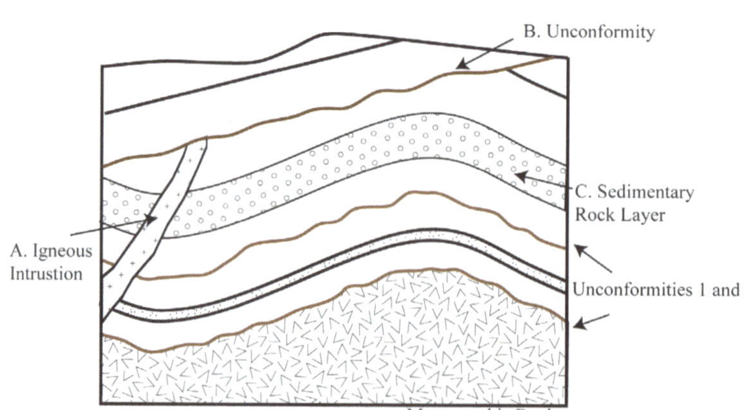

Figure 1.9 | Block diagram to use to answer questions 3, 4, and 5. Unconformities are shown in brown.

3. In Figure 1.9, which of the following geologic structures is the **youngest**?

   a. A b. B c. C

4. Which Geologic Law did you use to come to the conclusion you made in the previous question?

   a. The Law of Superposition

   b. The Law of Cross-Cutting

   c. The Law of Original Horizontality

   d. Unconformities

5. Examine unconformities 1 and 2 indicated in Figure 1.9. Which of the following statements about them is true?

   a. The older unconformity is a Nonconformity, while the younger is an Angular Unconformity.

   b. The older unconformity is a Disconformity, while the younger is a Nonconformity.

   c. The older unconformity is a Nonconformity, while the younger is a Disconformity.

   d. The older unconformity is an Angular Unconformity, while the younger is a Disconformity.

6. Examine the Unconformity shown in Figure 1.10. What type of unconformity is this?

   a. Angular Unconformity

   b. Nonconformity

   c. Disconformity

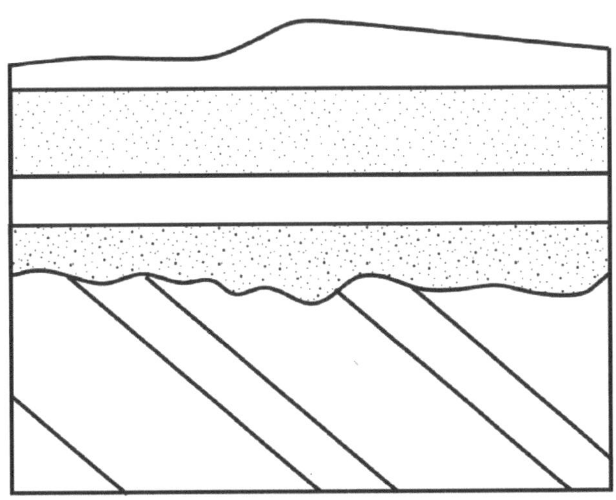

Figure 1.10 | Block diagram to use to answer question 6.

Examine Figure 1.11. Note that all of the layers in this block diagram are composed of sedimentary rock and the unconformities are colored in red. Using the geologic laws discussed earlier and following the examples shown in Figures 1.6 and 1.7, identify the geologic events that occurred in this area. Then place the following geologic events in the correct relative time sequence.

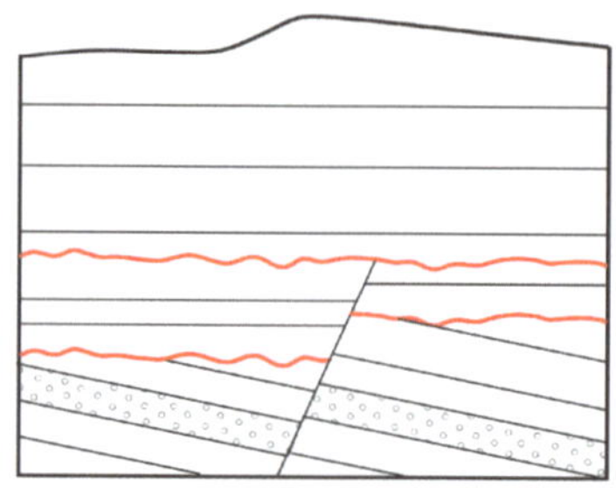

Figure 1.11 | Block diagram to use to answer questions 7, 8, and 9. Unconformities are shown in red.

a. Tilting.
b. Uplift and Erosion (Angular Unconformity).
c. Submergence and deposition of sedimentary layers 10-13.
d. Uplift and Erosion to current position.
e. Submergence and Deposition of sedimentary layers 7-9.
f. Uplift and Erosion (Disconformity)
g. Submergence and deposition of sedimentary layers 1-6.
h. Fault.

___	___	___	___	___	___	___	___
1 Oldest	2	3	4	5	6	7	8 Youngest

7. Which of the above geologic events is the second in the sequence?

   a. A b. B c. C d. D
   e. E f. F g. G h. H

8. Which of the above geologic events is the fifth in the sequence?

   a. A b. B c. C d. D
   e. E f. F g. G h. H

9. Which of the above geologic events is the seventh in the sequence?

   a. A b. B c. C d. D
   e. E f. F g. G h. H

1.3 FAUNAL SUCCESSION AND INDEX FOSSILS

Fossils serve as another valuable instrument in relative dating. Fossils are the preserved remnants of prehistoric species typically located within sedimentary rocks. Organisms emerge at distinct intervals in geological history and become extinct at disparate times. These creatures also undergo morphological changes over time. The pattern of emergence, transformation, and extinction of numerous fossil creatures establishes a discernible record of organisms maintained throughout geological time. Consequently, rocks of identical age probably have analogous fossils, which can be utilized to date sedimentary formations. This principle is referred to as the Law of Faunal Succession.

Certain fossils, known as Index Fossils, are very valuable for chronological dating. These species are likely to be discovered because to their historical abundance and propensity for fossilization, often attributed to possessing a sturdy skeletal structure. These species typically possess an extensive geographic distribution, allowing them to serve as indicator fossils across many regions. Nevertheless, they should possess a limited geologic range (the duration an organism exists on Earth), enabling us to ascertain the rock's age more accurately upon discovering the fossil. Index fossils are frequently the most expedient and straightforward method for precisely and correctly dating sedimentary strata.

1.3.1 LAB EXERCISE

Part B – Faunal Succession

The use of animals and their preserved remains (fossils) can help build a highly precise time sequence, often with a higher resolution than absolute dating. In the following section, use this principle to answer the following questions.

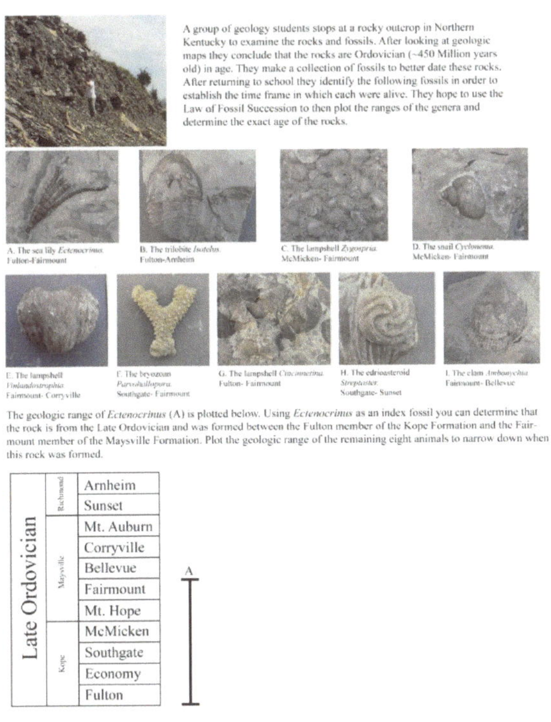

Figure 1.12

10. Based on the assemblage of organisms (A-I) in this sample, what is the age of this rock?

    a. Economy b. Southgate c. McMicken d. Mt. Hope

    e. Fairmount f. Bellevue g. Corryville h. Mt. Auburn

11. Which organism was the *most* useful in coming to this conclusion (which is the best index fossil)?

    a. Isotelus b. Zygospira c. Cyclonema d. Vinlandostrophia

    e. Parvohallopora f. Cincinnetina g. Streptaster h. Ambonychia

12. Which organism was the *least* useful in coming to this conclusion (which is the worst index fossil)?

    a. Isotelus b. Zygospira c. Cyclonema d. Vinlandostrophia

    e. Parvohallopora f. Cincinnetina g. Streptaster h. Ambonychia

1.4 ABSOLUTE TIME AND RADIOMETRIC DATING

Absolute time is a technique for ascertaining the age of a rock or object, mostly through the use of radiometric isotopes. Atoms consist of three fundamental particles: protons, electrons, and neutrons. All three particles are crucial to geological study: the proton count determines a specific element, the electron count governs the element's bonding capabilities to form compounds, and the neutron count alters the atomic weight of the element. Isotopes are atoms of an element that differ in the number of neutrons in their nucleus, resulting in variations in their atomic weight. An element with an excess or deficiency of neutrons in its nucleus becomes unstable and undergoes degradation over time, a process known as radioactive decay. Radioactive decay entails the emission of a particle from a radioactive atom, referred to as the parent atom, resulting in the transformation into another element, known as the daughter atom. We can analyze and quantify the radioactivity of various elements in the laboratory and compute the decay rate. Although the decay rates of isotopes range from milliseconds to billions of years, all radiometric isotopes undergo decay in a comparable manner. Radiometric decay adheres to a curve characterized by the half-life of a radiometric isotope. The half-life is the duration required for half of the atoms of a radiometric parent isotope to decay into the daughter isotope. The half-life is unaffected by the quantity of atoms present at any moment; hence, the duration required to decrease from 100% to 50% of the parent isotope is identical to that needed to decline from 50% to 25% of the parent isotope. By knowing the half-life of a specific radiometric isotope and measuring the quantities of both parent and daughter isotopes in a rock, we may determine the rock's age, a process known as Radiometric Dating. Considering the characteristics of the decay curve, a substance never completely depletes the parent isotope; yet, we can only effectively measure the parent isotope up to 10-15 half-lives.

1.4.1 LAB EXERCISE

Part C Radiometric Dating

Complete the following chart by calculating the amount of parent isotope remain ing for all of the given half-lives, then plot your findings on the graph (Figure 1.13). Make sure you connect the data points on the graph by drawing in the decay curve.

Use the completed chart and graph to answer the questions below.

Number of Half-lives	Amount of Parent Isotope Remaing
0	100%
1	50%
2	
3	
4	
5	
6	
7	
8	

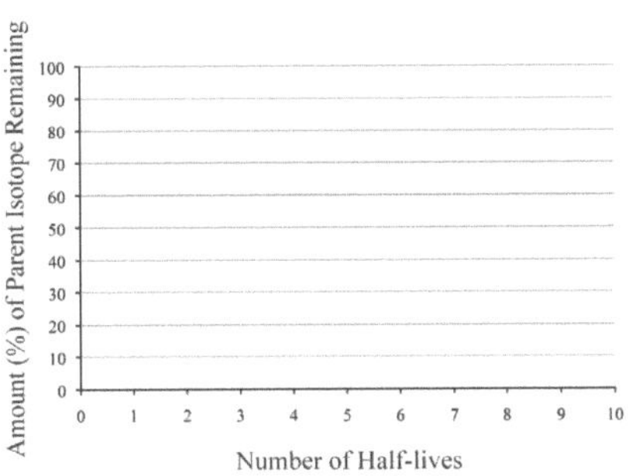

Figure 1.13

13. How much of the parent isotope would be remaining after 7 half-lives have passed?
    a. 6.25% b. 1.56% c. 0.78% d. 0.39%
14. If a radiometric element has a half-life of 425 years, how old would a rock be that only had 3.125% of the parent isotope remaining?
    a. 2125 years b. 1700 years c. 2550 years d. 3400 years
15. Based on your graph above, approximately how much of the parent isotope would be remaining after 3.5 half-lives?
    a. 16% b. 12% c. 4% d. 8%
16. Based on your graph above, approximately how many half-lives have passed when only 35% of the parent isotope is remaining?
    a 0.75 b. 1.5 c. 2.1 d. 2.5

1.5 DATING SYSTEMS

Numerous radioactive isotopes are frequently employed in absolute dating. Each of these systems serves distinct purposes in geology, necessitating various materials and enabling the dating of items within specified temporal parameters.

Carbon-14 dating has restricted applicability in geology, yet it remains the most recognized method among the general populace. Carbon-14, the parent isotope, is present in organic materials such as bone, tissue, plants, and fiber. This isotope occurs naturally in trace quantities in the atmosphere as part of CO_2 and is assimilated by plants during photosynthesis, thus permeating the food chain.

Your body presently has Carbon-14, which is degrading into Nitrogen-14 (the daughter isotope), yet you replenish it with each meal. When an organism ceases to consume food or a plant halts photosynthesis, the radioactive carbon begins to decay without replenishment, a process that can be readily quantified. Carbon-14 possesses a brief half-life of 5,730 years and is applicable for dating samples up to roughly 70,000 years old. The Earth's age is 4.54 billion years, hence carbon-14 is applicable solely for dating modern materials.

Uranium dating entails a sophisticated system of several isotopes that undergo decay in a chain reaction until they culminate in non-radiogenic lead. Uranium is unexpectedly present in numerous locations, albeit typically in extremely small quantities. A further complication with this technique is that the daughter isotope, lead, occurs naturally at various locations, complicating the differentiation between lead produced from radiometric decay and lead present in the environment. The mineral zircon addresses these difficulties by concentrating uranium and eliminating lead from its structure. Consequently, we employ Uranium dating on zircons located within igneous rocks, including volcanic ash or rocks generated at considerable depths in the Earth. Uranium possesses an extensive half-life of 4.5 billion years, sufficiently lengthy to date the majority of terrestrial rocks. The complicated system requires around one million years to stabilize, rendering Uranium dates shorter than this dubious.

Potassium-Argon dating is an effective technique for dating geological formations. Potassium decays into two distinct daughter isotopes: Argon and Calcium. We quantify the concentration of Argon in the rocks because, unlike calcium, it is infrequent in minerals due to its nature as a Noble Gas, which typically does not form bonds with other elements. Consequently, any argon present in a mineral originates from the decay of potassium. The utilization of Argon presents several disadvantages; for example, the gas might readily dissipate from a rock, necessitating meticulous precautions in the laboratory to avert this occurrence. This technique functions effectively when analyzing various materials high in potassium, such as granite, which is replete with potassium-rich pink minerals known as feldspars. The half-life of potassium is 702 million years, rendering it comparable to uranium in its efficacy for dating ancient rocks.

Despite the utilization of these technologies, the potential for mistake persists, necessitating the consideration of any specific radiometric date as a scientific hypothesis requiring further validation. Errors may arise from the addition or deletion of either parent or daughter isotopes in the rock post-formation. This can be accomplished in several methods, predominantly through the application of heat and pressure (metamorphism). Methods exist to rectify these difficulties, enabling the scientist to date both the rock and the metamorphic event, provided the geological history is understood.

It is uncommon to discover a rock that possesses a quantity of the parent isotope remaining that precisely coincides with one of the half-lives. Typically, a straightforward formula is required to determine a rock's age by utilizing the half-life duration and the quantity of surviving parent isotopes.

The formula is:

$$Age = -(\frac{t_{1/2}}{0.693})\ln(P)$$

t1/ 2 = The length of the half-life in years

P = The amount of the parent remaining in decimal form. For example, if there is 50% of the parent remaining it would equal 0.5.

Let's work an example using the equation that we already know the answer to in advance. You have a sample of bone that has 25% of the Carbon-14 (Half-life= 5730 years) remaining, how old is the sample? We can answer this question in two ways:

1. We know that if there is 25% remaining, two half-lives have passed and with each half-life being 5730 the bone would be 11,460 years old.
2. We could use the above equation and insert both the length of the halflife and the amount of the parent remaining:

$$Age = -(\frac{5730}{0.693})\ln(0.25)$$

To solve the equation, take the Natural Log (ln) of 0.25 and multiply by the term in the parentheses (make sure to include the negative sign). If you do this you will get 11,460 as well.

1.5.1 LAB EXERCISE

Part D – Isotopic Systems

Using what you learned in the previous section regarding absolute dating, determine the most appropriate methods and the ages of the materials in the following questions.

17. An Archeologist finds some cotton cloth at a burial site and wants to determine the age of the remains. Which isotopic system should they use?

    a. Carbon-14 b. Uranium c. Potassium-Argon

18. The Archeologist determines that there is 16.7% of the parent isotope remaining in the cloth sample. How old is the burial site? Hint: you can find the length of the half-life in the reading above.

    a.13,559 years b. 14,798 years

c.16,743 years
d. 1.66 billion years
e.1.81 billion years
f. 2.05 billion

19. A geologist is trying to date a sequence of sedimentary rocks with abundant fossils and sandstones. Within the sequence is a distinctive clay layer that under closer inspection is fine-grained volcanic ash. Which of the following is the best way to obtain an absolute date for the sequence of rocks?
    a.Carbon date the fossils
    b.Potassium-Argon date the sands
    c.Uranium date the Zircons in the ash
    d.Identify the index fossils
20. The geologist determines there is 78.3% of the parent remaining in the sample that they examine. How old is the sequence of rocks? Hint: you can find the length of the half-life in the reading above.
    a. 187.5 million years
    b. 247.8 million years
    c. 390.7 million years
    d. 2.504 billion years
    e. 1.588 billion years
    f. 1.202 billion years

1.6 STUDENT RESPONSES

The following is a summary of the questions in this lab for ease in submitting answers online.

1. In Figure 1.8, which of the following rock layers is **oldest**?
   a. A b. B c. C
2. Which Geologic Law did you use to come to the conclusion you made in the previous question?
   a.The Law of Superposition
   b.The Law of Cross-Cutting
   c.The Law of Original Horizontality
   d.Unconformities
3. In Figure 1.9, which of the following geologic structures is **youngest**?
   a. A b. B c C
4. Which Geologic Law did you use to come to the conclusion you made in the previous question?
   a.The Law of Superposition
   b.The Law of Cross-Cutting

c. The Law of Original Horizontality

d. Unconformities

5. Examine unconformities 1 and 2 indicated in Figure 1.9. Which of the following statements about them is true?

   a. The older unconformity is a Nonconformity, while the younger is an Angular Unconformity.

   b. The older unconformity is a Disconformity, while the younger is a Nonconformity.

   c. The older unconformity is a Nonconformity, while the younger is a Disconformity.

   d. The older unconformity is an Angular Unconformity, while the younger is a Disconformity.

6. Examine the Unconformity shown in Figure 1.10. What type of unconformity is this?

   a. Angular Unconformity b. Nonconformity c. Disconformity

7. Which of the above geologic events is the second in the sequence?

   a. A b. B c. C d. D

   e. E f. F g. G h. H

8. Which of the above geologic events is the fifth in the sequence?

   a. A b. B c. C d. D

   e. E f. F g. G h. H

9. Which of the above geologic events is the seventh in the sequence?

   a. A b. B c. C d. D

   e. E f. F g. G h. H

10. Based on the assemblage of organisms (A-I) in this sample, what is the age of this rock?

    a. Economy b. Southgate c. McMicken d. Mt. Hope

    e. Fairmount f. Bellevue g. Corryville h. Mt. Auburn

11. Which organism was the *most* useful in coming to this conclusion (which is the best index fossil)?

    a. Isotelus b. Zygospira c. Cyclonema d. Vinlandostrophia

    e. Parvohallopora f. Cincinnetina g. Streptaster h. Ambonychia

12. Which organism was the *least* useful in coming to this conclusion (which is the worst index fossil)?

    a. Isotelus b. Zygospira c. Cyclonema d. Vinlandostrophia

    e. Parvohallopora f. Cincinnetina g. Streptaster h. Ambonychia

13. How much of the parent isotope would be remaining after 7 half-lives have passed?

    a. 6.25% b. 1.56% c. 0.78% d. 0.39%

14. If a radiometric element has a half-life of 425 years, how old would a rock be that only had 3.125% of the parent isotope remaining?

    a. 2125 years b. 1700 years c. 2550 years d. 3400 years

15. Based on your graph above, approximately how much of the parent isotope would be remaining after 3.5 half-lives?

    a. 16% b. 12% c. 4% d. 8%

16. Based on your graph above, approximately how many half-lives have passed when only 35% of the parent isotope is remaining?

    a. 0.75 b. 1.5 c. 2.1 d. 2.5

17. An Archeologist finds some cotton cloth at a burial site and wants to determine the age of the remains. Which isotopic system should they use?

    a. Carbon-14 b. Uranium c Potassium-Argon

18. The Archeologist determines that there is 16.7% of the parent isotope remaining in the cloth sample. How old is the burial site? Hint: you can find the length of the half-life in the reading above.

    a.13,559 years b. 14,798 years

    c.16,743 years d. 1.66 billion years

    e.1.81 billion years f. 2.05 billion

19. A geologist is trying to date a sequence of sedimentary rocks with abundant fossils and sandstones. Within the sequence is a distinctive clay layer that under closer inspection is fine-grained volcanic ash. Which of the following is the best way to obtain an absolute date for the sequence of rocks?

    a Carbon date the fossils

    b. Potassium-Argon date the sands

    c. Uranium date the Zircons in the ash

    d. Identify the index fossils

20. The geologist determines there is 78.3% of the parent remaining in the sample that they examine. How old is the sequence of rocks? Hint: you can find the length of the half-life in the reading above.

    a.187.5 million years b. 247.8 million years

    c.390.7 million years d. 2.504 billion years

    e. 1.588 billion years f. 1.202 billion years

2 Chapter

INTERIOR OF THE EARTH

2.1 INTRODUCTION: AN OVERVIEW

Investigating the Earth's interior presents a considerable challenge owing to the absence of direct access. Numerous activities occurring at the Earth's surface are propelled by the heat produced within the Earth, therefore becoming a grasp of its interior crucial. Volcanism, seismic activity, and several terrestrial phenomena are consequences of internal processes occurring within the Earth.

A significant portion of our understanding of the Earth's interior is derived from indirect methods, particularly the analysis of seismic data to ascertain its underlying structure. In the early 1900s, scientists realized that seismic waves produced by earthquakes could be utilized to differentiate the characteristics of the Earth's interior layers. The velocity of primary and secondary waves (P and S waves) varies according to the density of the materials they traverse. Consequently, seismic waves do not propagate through the Earth in linear paths; instead, they are reflected and refracted, signifying that the Earth's composition is not uniform.

The Earth's interior comprises an inner core, outer core, mantle, and crust. The inner core, situated near the Earth's center, is highly dense, subjected to immense pressure, and is believed to consist of an iron-nickel alloy. It is solid and encircled by a liquid iron and nickel region known as the outer core. The outer core is believed to generate the Earth's magnetic field. A substantial fraction of the Earth's volume is in the mantle, which encases the core. This layer is less dense than the core and comprises a solid that can exhibit plastic (deformable) behavior. The Earth's outermost layer is the crust. The two varieties, continental and oceanic crust, differ in thickness, composition, and density.

2.2 INTERIOR OF THE EARTH

The analysis of seismic waves and their propagation through the Earth has been helpful in identifying variations in density and composition within the Earth, as well as in delineating the boundaries between the inner core, outer core, mantle, and crust. Seismic waves are energy waves produced during earthquakes; two varieties, P and S waves, spread through the Earth as wave fronts from their source. P-waves are compressional waves that oscillate longitudinally, like to an accordion, whereas S-waves are shear waves that displace material perpendicularly to their direction of propagation, resembling the action of snapping a rope. The velocity of both waves escalates with the increasing density of the materials they traverse. Due to the fact that most liquids possess a lower density than their solid equivalents, and as seismic velocity is contingent upon density, the existence of any liquid phase within the Earth's interior will influence seismic waves. S waves cannot propagate through liquids, as their lateral motion cannot be sustained in fluids; thus, it is understood that the outer core is liquid.

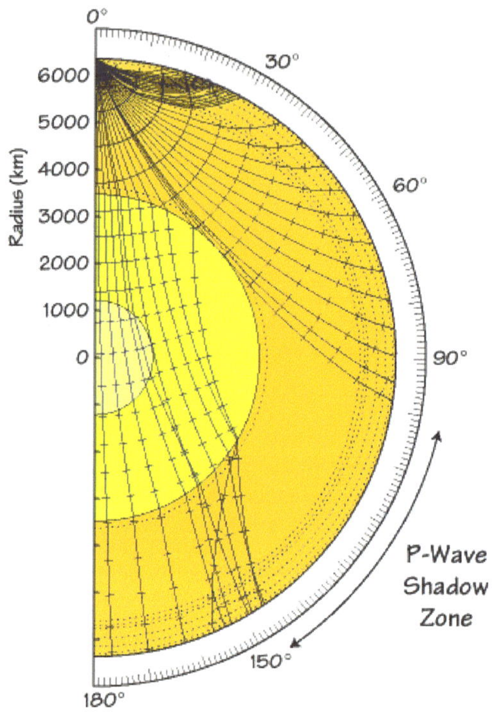

Figure 2.1 | A depiction of the P-wave shadow zone.

If the Earth was completely homogeneous, the P and S waves would flow in straight lines. They don't behave this way, however. As the waves travel through materials of different densities, they are refracted, or bent, as their direction and velocity alter. Sometimes these refractions can result in shadow zones, which are areas along the Earth where no seismic waves are detected. Due to the presence of a liquid outer core, a P-wave shadow zone exists from 103o-143o (see Figure 2.1) from the earthquake origination point (focus), and a larger S-wave shadow zone exists in areas greater than 103o from the earthquake focus.

Based on the way that the Earth travels through space, we know that the average density of the Earth is 5.52 g/cm3. When rocks at the Earth's surface are analyzed, we find that most crustal rocks have densities in the range of 2.5-3 g/cm3, which is lower than the Earth's average. This means that there must be denser material inside the Earth to arrive at that higher average density; in fact, the core region of the Earth is estimated to have a density of 9-13 g/cm3. The composition of the Earth's layers also changes with depth. The bulk Earth composition is mostly made up of iron (~32%), oxygen (~30%), silicon (~16%), and magnesium (15%). If you examine rocks at the Earth's surface, however, you will find that oxygen is the most abundant element by far (~47%), followed by silicon (~28%) and aluminum (~8%), and lesser amounts of iron, calcium, sodium, potassium, and magnesium. Minerals made from silicon and oxygen are very important and are called silicates. So, if iron is present in lower numbers in the crustal rocks, where has that iron gone? Much of it can be found in the core of the Earth, which accounts for the major increase in density there. Review Table 2.1 below for general information about each layer of the Earth, and note how much thicker continental crust is compared to oceanic crust. Examine Figure 2.2 for a depiction of the layers of the Earth.

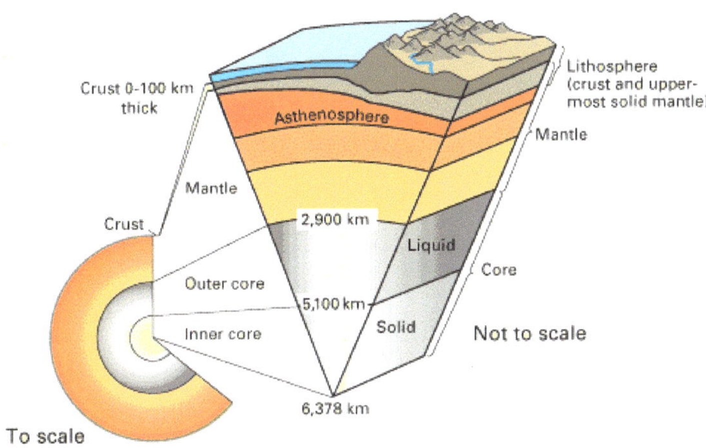

Figure 2.2 | A depiction of the inner layers of the Earth. Note that the image to the bottom left is to scale, while the image to the right is not.

Note that this figure includes the terms lithosphere and asthenosphere. The lithosphere is the outer, rigid part of the Earth made up of the upper mantle, oceanic crust, and continental crust. The asthenosphere is just beneath the lithosphere and, rather than being rigid, behaves plastically and flows.

Table 2.1

Earth's Layer	Density (g/cm3)	Thickness (km)	Composition
Continental Crust	~2.7-2.9	~20-70	Felsic rocks
Oceanic Crust	~3.0	~8-10	Mafic rocks
Mantle	~3.4-5.6	~2,885	Ultramafic rocks
Outer Core	~9.9-12.2	~2,200	Iron, some sulfur, nickel, oxygen, silicon
Inner Core	~12.8-13.1	~1,220	Iron, some sulfur and nickel

Relatively recent advances in imaging technology have been used to better understand the Earth's interior. **Seismic tomography** has been used to give a more detailed model of the Earth's interior. In CAT scans, x-rays are aimed at a person and rapidly rotated, generating cross-sectional images of the body. In a similar fashion, repeated scans of seismic waves are stacked to produce a three-dimensional image in seismic tomography. This technique has been used in many ways, from searching for petroleum near the Earth's surface to imaging the planet as a whole. Figure 2.3 depicts an image of the mantle created from seismic tomography.

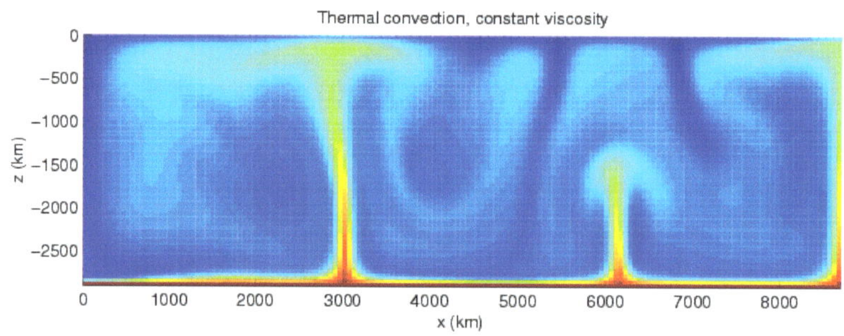

Figure 2.3 | A model of thermal convection in the mantle, created using seismic tomography. This model depicts areas of cool mantle material in blue and areas of warm mantle material in red. The thin red areas represent rising plumes.

2.2.1 LAB EXERCISE

Part A – Interior of the Earth

The following graph (Figure 2.4) displays seismic velocities (in kilometers per second) of P and S waves with depth (measured in kilometers) inside the Earth. Examine the graph closely and answer the following questions.

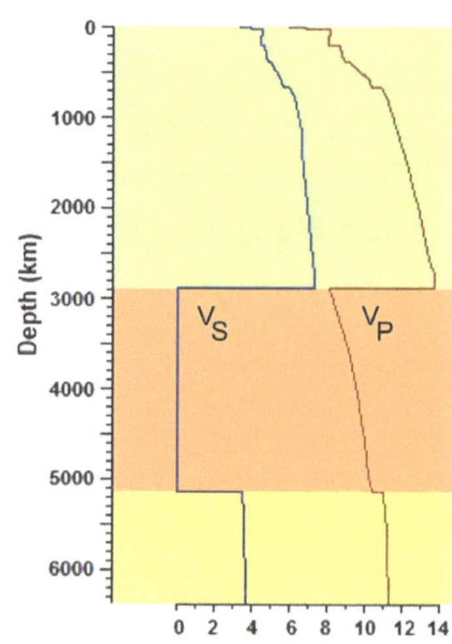

Figure 2.4 | Graph of seismic velocity with depth within the Earth's Interior.

1. Observe the velocities of the waves on the graph. Which one travels faster?

   a. P waves b. S waves

2. Inspect the P wave velocities. Where do the P wave velocities abruptly change?

   a. ~20 km b. ~2,900 km c. ~5,100 km d. All of the above

3. In which zones do the P wave velocities appear to be steadily increasing?

   a.~20–2,900 km b. ~2,900-5,100 km

   c.~5,100-6,400 km d Both a & b

4. Observe the S wave velocities. Where do the S wave velocities abruptly change?

   a. ~20 km b. ~2,900 km c. ~5,100 km d. All of the above

5. At ~2,900 km, the S wave velocity falls to 0. Why?

   a.S waves can't travel through solids, and this depth is where the solid inner core exists.

   b.S waves can't travel through liquids, and this depth is where the liquid outer core exists.

   c.S waves can't travel through solids, and this depth is where the solid mantle exists.

   d.S waves entered the shadow zone.

Observe closely the changes in seismic wave velocity. You may add lines to your graph to denote the abrupt changes. Label each zone with the internal layers of the Earth and answer the following questions.

6. The zone from ~0-20 km represents the Earth's:

   a. crust b. mantle c. inner core d. outer core

7. The zone from ~20-2,900 km represents the Earth's:

   a. crust b. mantle c. inner core d. outer core

8. The zone from ~2,900-5,100 km represents the Earth's:

   a. crust b. mantle c. inner core d. outer core

9. The zone from ~5,100-6,400 km represents the Earth's:

   a. crust b. mantle c inner core d. outer core

2.3 EARTH'S MAGNETIC FIELD

The examination of seismic waves and their transmission through the Earth has facilitated the identification of density and compositional differences, as well as the delineation of boundaries between the inner core, outer core, mantle, and crust. Seismic waves are energy waves generated by earthquakes; two types, P and S waves, propagate through the Earth as wave fronts from their origin. P-waves are compressional waves that oscillate longitudinally, akin to an accordion, whereas S-waves are shear waves that displace material perpendicularly to their direction of propagation, simulating the action of snapping a rope. The speed of both waves increases with the rising density of the materials they pass through. Because most liquids have a lower density than their solid counterparts, and seismic velocity depends on density, the presence of any liquid phase within the Earth's interior will affect seismic waves. S waves are unable to travel through liquids due to their lateral motion being unsustainable in fluids; hence, it is inferred that the outer core is in a liquid state.)

Figure 2.5 | Solar wind interacting with the Earth's magnetic field.

The magnetic field undergoes continuous modifications and has undergone multiple polarity reversals in the past, albeit these reversals remain poorly comprehended. The examination of historical reversals depends on paleomagnetism, which is the record of residual magnetism retained in specific rock types. Iron-bearing crystals originating from lava can orient themselves with the Earth's magnetic field, so preserving a historical record of the planet's magnetic field. Nonetheless, this retained magnetism may be forfeited if the mineral within the rocks has not been subjected to temperatures over the Curie point (the threshold above which minerals relinquish their magnetic). The iron atoms effectively "secure" themselves in alignment with the magnetic pole. This documents the orientation of the magnetic field at that time (we are presently in a normal polarity, where north on a compass corresponds roughly with geographic north, or the North Pole). If the magnetic field were stationary, all magnetic minerals would align in the same direction. This is not the situation, however. Reversals transpire with considerable frequency on the geological time scale.

Figure 2.6 | An example of the beautiful Aurora Borealis, light displays created by solar storm interaction with the Earth's magnetic field.

Magnetic poles not only reverse throughout geological time but also exhibit movement. Paleomagnetic studies indicate that the magnetic poles exhibit regular movement, traversing the globe. Polar wandering curves illustrate the movement of the poles over the Earth's surface throughout time. Apparent polar wander denotes the observed displacement of the Earth's paleomagnetic poles in relation to a stationary continent (Figure 2.7). In the Plate Tectonics chapter, you will discover that polar wandering curves provide as compelling evidence for the hypothesis of plate movement, as the curves for various continents do not align with the locations of the magnetic poles. They all converge on the present location of the pole; nonetheless.

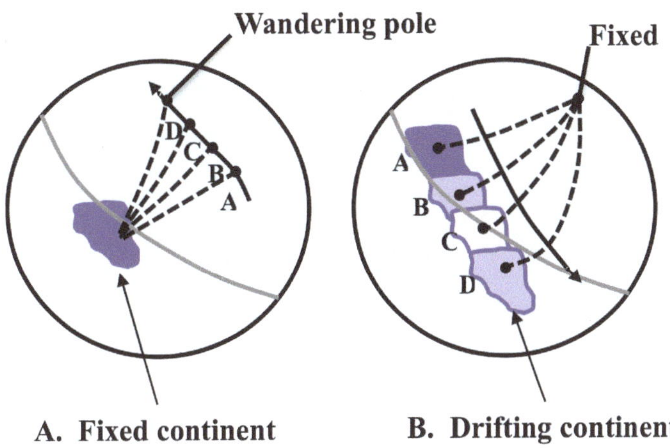

Figure 2.7 | If continents are fixed, as in A in the figure, the pole must be wandering. However, the pole is relatively fixed around the pole (with some movement), so the drifting continent (B) is the correct model.

2.3.1 LAB EXERCISE

Part B – Earth's Magnetic Field

You will use the polar map given (Figure 2.8) to plot the changing locations of the magnetic pole over time. To view the polar map, imagine that you are above the North Pole looking down on it. 90o N latitude is directly in the center of the map, and the lines of latitude, measured in 2 degree increments, spread out in circles from the center. Values of longitude are also given, and are represented as lines that radiate out from the center in increments of 30 degrees. A scale bar, in kilometers, is provided. To familiarize yourself with the map, first practice plotting some of the locations given. The first two, A and B, have been done for you. You will be plotting actual locations of the magnetic North Pole over time ranging from 14001900 AD onto Figure 2.8. The table below includes the year and latitude and longitude (as degrees from 0 to 360) for each location. Once you have completed this, answer the questions that follow. Remember that 1,000 meters = 1 kilometer.

Point	Year	Latitude	Longitude
A		87.7	160.3
B		85.6	6.6
C	1400	84.8	228.3
D	1500	86.3	301.5
E	1600	85.6	316.7
F	1700	81.1	307.1
G	1800	81.1	297.1
H	1900	82.3	288.2

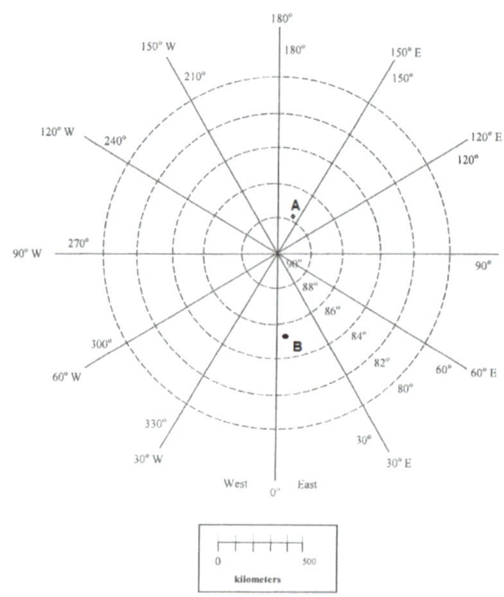

Figure 2.8 | Add the magnetic pole locations to this polar map and answer the questions below.

10. Measure the distance, using the bar scale, between the pole in 1400 and 1500 (locations C and D). How far did the pole move?
    a ~50 km b ~150 km c. ~600 km d. ~1,000 km

11. How far (in km) did the pole move in one year during this time period?
    a. 0.5 km b. 1.5 km c. 6 km d. 10 km

12. How far did the pole move in meters in one year during this time period?
    a. 1,500 m b. 500 m c. 10,000 m d. 6,000 m

13. Approximately how far did the pole move per day?
    a. <1 m b. ~16 m c. ~20 m d. ~12 m

14. Measure the distance, using the bar scale, between the pole in 1400 and 1700 (locations C and F). How far did the pole move?

    a. ~50 km b. ~150 km c. ~600 km d. ~1,000 km

15. Do pole movements tend to be steady every 100 years or variable?

    a. movements are steady b. movements are variable

2.4 GOOGLE EARTH INTRODUCTION

Google Earth is an excellent resource for visualizing and examining numerous geological aspects that will be addressed in this course. This program is complimentary and user-friendly. This initial Google Earth assignment will concentrate on acclimating you to the program and certain technologies that will be utilized in subsequent labs. The most effective method to see geology is to venture outdoors. As an online class is not feasible, the alternative is utilizing Google Earth. This program is pragmatic and beneficial, with numerous uses.

NOTICE: Google Earth updates versions periodically. If this occurs the instructions in the labs for this chapter may refer to the older version of the program. If you think this has occurred please let your instructor know.

If Google Earth is not already installed on the computer you are using**, the** please do the following:

1. Go to http://earth.google.com
2. Click on the Download Google Earth tab at the top of the page, **review** the Privacy Policy, and click Agree and Download to download the latest version.
3. Save the file to **your** desktop**,** open it and follow the instructions to install.
4. Open Google Earth.

Before we begin the assignment, let's first familiarize ourselves with Google Earth. Read each step and spend a few minutes trying things out, which will make things easier later. Also note that the Mac and PC versions of Google Earth are a little diff

You will use this program extensively throughout the course. Take the time to learn how to navigate it now.

Step 1 – Navigation

Watch each of the tutorial videos at**:** http://www.google.com/earth/learn/beginner.html#navigation. It is important that you dedicate time to review these virtual resources to help you better **understand Google Earth and its capabilities**. These will be key to mastering the tool. Navigating in Google Earth can be done in two ways:

First, you can use the Search panel in the upper left **hand side of the screen.** Just type in a location, address, or coordinate and it will zoom into the position (give it a try **now**).

The second approach is that you can also navigate manually:

- To move position you can left click with the mouse and drag the map or click on the hand icon in the upper right corner.
- You can zoom in and out using the mouse wheel, by right clicking and dragging the mouse up or down, or by sliding the lowest bar in the upper right corner.
- Click and hold the mouse wheel in order to rotate the map (left and right) or tilt the scenery (up and down). This can also be done using the arrows surrounding the eye icon in the upper right corner.

Along the bottom of the image it gives several important pieces of information:

1. Latitude and Longitude
2. Elevation in reference to sea level
3. Eye altitude, which indicates how zoomed in or out you are.

For example, let's check out Niagara Falls. In the Search panel in the upper left, type in Niagara Falls, NY. To better tell where the Falls are, you want to zoom out a little bit. Notice your eye altitude along the bottom right. Use the minus button for the zoom until you are at ~10,000' eye altitude. Check your latitude (~43o04'39"N – read as 43 degrees, 4 minutes, and 57 seconds North) and longitude (~79o04'28"W). It will move as you move your cursor across the screen, as will your elevation. If you want to see a picture of the Falls, just click on one of the many photo icons to see one. Note that the river is headed in a general northerly route – you can zoom in close to see the actual Falls.

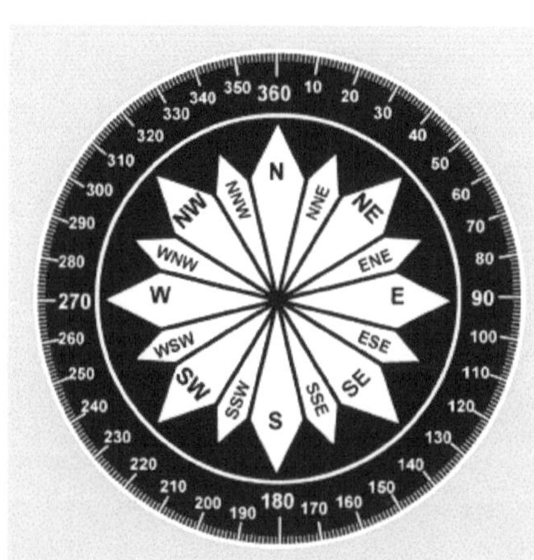

Figure 2.9 | A compass displaying the degrees.

Also important to understand with navigation is the concept of bearing. A bearing is the compass direction as measured between two points. It can be expressed as an azimuth bearing in degrees between 0 and 360, as along a circle. 0 and 360 degrees would be north, 90 degrees would be east, south would be 180 degrees, and west would be 270 degrees (Figure 2.9).

Bearing can be useful in determining how to get from point A to point B. To measure a bearing in Google Earth, follow this procedure. You want the bearing from point A to point B. Locate point A. Select the Ruler tool. Measure from A to B. In the last line in the Ruler tool box, read the Heading in degrees – that is measuring bearing (Tip: when point A is very far from point B, it can be helpful to add a placemark. Once at point A, choose the item in the menu bar right above the image that looks like a yellow pushpin. That is a placemark. Place it at point A. Also add a placemark at point B. Zoom out far enough that you can see both placemarks, then use the ruler tool to measure the heading between them. This works well for measuring headings across the entire United States). Let's practice getting a bearing again with Niagara Falls. For this example, point A will be the Scotiabank Convention Center located at 43o04'34.08"N and 79o05'05.98"W. Point B will be the name label for Goat Island (located across the river, just before the Falls begin). The bearing from point A to point B is ~74 degrees.

Step 2 – Measuring

To analyze features, we must be able to measure them, a task efficiently accomplished in Google Earth. The Ruler Tool is utilized for measurement, accessible by clicking the ruler icon in the toolbar above the image or by selecting Ruler from the Tools menu at the top.

The ruler tool has two options: line and path. The line option, which is the default setting, provides the distance and direction between two points; see the dropdown menu that offers 11 distinct units of measurement. To obtain a measurement, after selecting the ruler tool, you merely click on two distinct spots. The path option provides the distance between two or more sites, enabling the measurement of non-linear distances.

NOTE: When measuring features you want to use the Map Length – using the ground length can lead to an incorrect answer.

Gradient will often be measured for this and future labs. Gradient is similar to slope which indicates how steep or flat an area is. It is calculated as the difference in elevation divided by the horizontal distance. When calculating gradient, maintain the same units in the numerator and denominator.

Gradient = change in elevation/horizontal distance

Let's practice again at Niagara Falls. First move the image slightly higher to fully see the start of the white water, just before the Falls begin (do this by left clicking the mouse and using the hand to move the image). With the eye altitude still at ~10,000', let's measure the distance across the river right at the start of the whitewater before the Falls (where the whitewater stops). First click your ruler icon, then select a point on one side of the river, then move your mouse straight across to the other side. In feet, this should measure ~4,800 feet (don't stop at the island – measure all the way to the other river bank). Using the pull-down menu, you can change the feet to miles, and the result should be ~0.9 miles. Now let's practice gradient across the actual Falls. Position your cursor over the actual Falls, and zoom in to an eye altitude of ~1,000 feet. Hold your cursor over the top of the Falls and record the elevation (remember, this is located along the bottom bar). Now move your cursor to the bottom of the Falls and record the elevation. The change in elevation (highest-lowest) will be your numerator. Use the Ruler tool to measure the distance between the two places – this will be your horizontal distance (the denominator). There will be variation in this answer depending on your exact spot along the Falls, but results should be similar to this:

Gradient = (500' – 325') / 75' = 175'/75' = 2.3

Step 3 – Changing the Options

For a few tasks it will be important to change some of the default settings on Google Earth in order to see a feature better or make your work easier. These changes can all be made by going to Tools in the menu bar across the top, then Options in the PC version (for the MAC, go to Google Earth, then Preferences).

1. Changing the unit for Elevation – From the 3D view tab, in the middle of the box there is a section entitled "Units of Measurement" that you can change between metric and English units.
2. Exaggerating Features – Since differences in elevations are much smaller than geographic distances it is sometimes hard to see features. To exaggerate features (that is, make a mountain look taller than it actually is in order to see it better), click on the 3D view tab, in the lower left side at the "Terrain" section of the box, look for "Elevation Exaggeration (also scales 3D buildings and trees)". If you want to exaggerate a feature increase this value up to 3. To view the area without any exaggeration, return the value to the default of 1.

2.4.1 LAB EXERCISE

Part D – Google Earth

Answer the following questions using the skills discussed above. In order to prepare you for examining geologic features in Google Earth, let's first examine a more familiar area, Washington, D.C.

16. Search for the "Washington Monument" and zoom into ~3,000 feet eye altitude. What is the latitude of the monument?

    a. 38o 53'N b. 38 o 53'S c. 77 o 02'E d. 77 o 02'W

17. What is the longitude of the monument?

    a. 38o 53'N b. 38 o 53'S c. 77 o 02'E d. 77 o 02'W

18. Zoom into an eye altitude of ~500 feet. Locate the base of the monument (the base is square and is located in the center of the circle). What is the elevation of the base of the monument in feet?

    a. 15 feet b. 40 feet c. 65 feet d. 80 feet

19. What is the elevation of the base of the monument in meters?

    a. 40 meters b. 25 meters c. 12 meters d. 7 meters

20. How big is the base (area) of the Washington Monument in square feet? (Hint – make sure the 3D Buildings is selected in the Layers box, located in the lower left of the screen. Also, area is measured in square feet, so make sure you are multiplying two measurements.)

    a.~1040 feet2 b. ~3100 feet2 c. ~4700 feet2 d. ~6030 feet2

21. If you were standing at the Washington Monument, what direction would you need to walk to go to the United States Capitol Building?

    a. North b. East c. South d. West

22. The direction system is useful, but imprecise. It is better to use a bearing. If you were standing at the Washington Monument, what bearing would you need to walk to go to the Smithsonian National Museum of Natural History, which houses many important geological specimens (Hint: Many of the buildings on the National Mall are part of the Smithsonian – make sure you get the correct building!)?

    a. 30o b. 75 o c. 90 o d. 260 o

23. If you decided instead to walk from the Washington Monument to the White House, how far would you have to walk, in miles (assume you could walk right to the entrance of the building)?

    a. 0.56 miles b. 0.78 miles c. 1.20 miles d. 1.87 miles

24. How far is this distance in kilometers?

    a. 0.5 kilometers b. 0.9 kilometers c. 1.4 kilometers d. 1.9 kilometers

25. Overall, would you be walking uphill or downhill?
    a. No change in elevation   b. Slightly uphill   c. Slightly downhill

26. How much does the elevation change in feet?
    a. 0 feet   b. 10 feet   c. 20 feet   d. 30 feet

27. What is the gradient of your walk?
    a. ~0.003   b. ~0.05   c. ~0.5   d. ~1

28. You decide to start walking from the Washington Monument to the White House, but as soon as you start a Park Ranger yells at you for walking on the grass. How far would the walk be (in miles) if you stayed on the sidewalks? There are many possible routes – try to take one of the shortest routes possible.
    a. 0.5 miles   b. 0.8 miles   c. 2 miles   d. 5 miles

29. Now, let's look at a geologic feature. Put 36 05 35.38 N 113 14 43.70 W into the search bar and zoom out to an eye altitude of ~25,000 feet. This is the Grand Canyon. If you started out in Washington, D.C., what bearing would you need to travel in to go to the Grand Canyon?
    a. 210o   b. 245o   c. 275o   d.315o

30. The Grand Canyon is an extraordinarily steep area. If you go from the river at the bottom of the Canyon (36 05 35.38 N 113 14 43.70 W) straight north up the canyon rim (the area that flattens out above the red layers of rock – 36 06 10.12 N 113 14 48.88 W), what is the gradient of the Grand Canyon? (**Hint:** You can tilt the image, which will make the flat rim easier to see, by either pressing down on the middle mouse wheel and moving the mouse forward and backward or by clicking on the arrows above and below the eye in the upper right hand portion of the window.)
    a. 0.214   b. 0.673   c. 0.976   d. 1.245

2.5 STUDENT RESPONSES

1. Observe the velocities of the waves on the graph. Which one travels faster?
   a. P waves   b. S waves

2. Inspect the P wave velocities. Where do the P wave velocities abruptly change?
   a. ~20 km   b. ~2,900 km   c. ~5,100 km   d. All of the above

3. In which zones do the P wave velocities appear to be steadily increasing?
   a.~20–2,900 km   b. ~2,900-5,100 km
   c.~5,100-6,400 km   d. Both a & b

4. Observe the S wave velocities. Where do the S wave velocities abruptly change?
   a. ~20 km   b. ~2,900 km   c ~5,100 km   d. All of the above

5. At ~2,900 km, the S wave velocity falls to 0. Why?

   a. S waves can't travel through solids, and this depth is where the solid inner core exists.

   b. S waves can't travel through liquids, and this depth is where the liquid outer core exists.

   c. S waves can't travel through solids, and this depth is where the solid mantle exists.

   d. S waves entered the shadow zone.

6. The zone from ~0-20 km represents the Earth's:

   a. crust   b. mantle   c. inner core   d. outer core

7. The zone from ~20-2,900 km represents the Earth's:

   a. crust   b. mantle   c. inner core   d. outer core

8. The zone from ~2,900-5,100 km represents the Earth's:

   a. crust   b. mantle   c. inner core   d. outer core

9. The zone from ~5,100-6,400 km represents the Earth's:

   a. crust   b. mantle   c. inner core   d. outer core

10. Measure the distance, using the bar scale, between the pole in 1400 and 1500 (locations C and D). How far did the pole move?

    a ~50 km   b. ~150 km   c. ~600 km   d. ~1,000 km

11. How far (in km) did the pole move in one year during this time period?

    a. 0.5 km   b. 1.5 km   c. 6 km   d. 10 km

12. How far did the pole move in meters in one year during this time period?

    a. 1,500 m   b. 500 m   c. 10,000 m   d. 6,000 m

13. Approximately how far did the pole move per day?

    a <1 m   b. ~16 m   c ~20 m   d. ~12 m

14. Measure the distance, using the bar scale, between the pole in 1400 and 1700 (locations C and F). How far did the pole move?

    a. ~50 km   b. ~150 km   c. ~600 km   d. ~1,000 km

15. Do pole movements tend to be steady every 100 years or variable?

    a. movements are steady   b. movements are variable

16. Search for the "Washington Monument" and zoom into ~3,000 feet eye altitude. What is the latitude of the monument?

    a. 38 53'N   b. 38 53'S   c. 77 02'E   d. 77 02'W

17. What is the longitude of the monument?

    a. 38 53'N   b. 38 53'S   c. 77 02'E   d. 77 02'W

18. Zoom into an eye altitude of ~500 feet. Locate the base of the monument (the base is square and is located in the center of the circle). What is the elevation of the base of the monument in feet?

    a. 15 feet   b. 40 feet   c. 65 feet   d. 80 feet

19. What is the elevation of the base of the monument in meters?

    a. 40 meters   b. 25 meters   c. 12 meters   d. 7 meters

20. How big is the base (area) of the Washington Monument in square feet? (Hint – make sure the 3D Buildings is selected in the Layers box, located in the lower left of the screen. Also, area is measured in square feet, so make sure you are multiplying two measurements.)

    a. ~1040 feet   b. ~3100 feet   c. ~4700 feet   d. ~6030 feet

21. If you were standing at the Washington Monument, what direction would you need to walk to go to the United States Capitol Building?

    a. North   b. East   c. South   d. West

22. The direction system is useful, but imprecise. It is better to use a bearing. If you were standing at the Washington Monument, what bearing would you need to walk to go to the Smithsonian National Museum of Natural History, which houses many important geological specimens (Hint: Many of the buildings on the National Mall are part of the Smithsonian – make sure you get the correct building!)?

    a. 30   b. 75   c. 90   d. 260

23. If you decided instead to walk from the Washington Monument to the White House, how far would you have to walk, in miles (assume you could walk right to the entrance of the building)?

    a. 0.56 miles   b. 0.78 miles   c. 1.20 miles   d. 1.87 miles

24. How far is this distance in kilometers?

    a. 0.5 kilometers   b. 0.9 kilometers   c. 1.4 kilometers   d. 1.9 kilometers

25. Overall, would you be walking uphill or downhill?

    a. No change in elevation   b. Slightly uphillc. Slightly downhill

26. How much does the elevation change in feet?

    a. 0 feet   b. 10 feet   c. 20 feet   d. 30 feet

27. What is the gradient of your walk?

    a. ~0.003   b. ~0.05   c. ~0.5   d. ~1

28. You decide to start walking from the Washington Monument to the White House, but as soon as you start a Park Ranger yells at you for walking on the grass. How far would the walk be (in miles) if you stayed on the sidewalks? There are many possible routes – try to take one of the shortest routes possible.

    a. 0.5 miles   b. 0.8 miles   c. 2 miles   d. 5 miles

29. Now, let's look at a geologic feature. Put 36 05 35.38 N 113 14 43.70 W into the search bar and zoom out to an eye altitude of ~25,000 feet. This is the Grand Canyon. If you started out in Washington, D.C., what bearing would you need to travel in to go to the Grand Canyon?

    a 210 b. 245 c. 275 d. 315

30. The Grand Canyon is an extraordinarily steep area. If you go from the river at the bottom of the Canyon (36 05 35.38 N 113 14 43.70 W) straight north up the canyon rim (the area that flattens out above the red layers of rock – 36 06 10.12 N 113 14 48.88 W), what is the gradient of the Grand Canyon? (**Hint:** You can tilt the image, which will make the flat rim easier to see, by either pressing down on the middle mouse wheel and moving the mouse forward and backward or by clicking on the arrows above and below the eye in the upper right hand portion of the window.)

    a. 0.214 b. 0.673 c. 0.976 d. 1.245

3 Chapter
TOPOGRAPHICAL MAPS

3.1 INTRODUCTION: AN OVERVIEW

A topographic map is a highly beneficial map that incorporates a third dimension (vertical) to a two-dimensional representation specified by the cardinal directions: north, south, east, and west. The third dimension of a topographic map is depicted by contour lines, which are hypothetical lines that indicate elevation above average sea level (a.s.l.) or mean sea level (m.s.l.). A map including elevation lines will furnish the reader with comprehensive information about the configuration of the Earth's surface. Understanding how to understand a topographic map enables an individual to locate and identify characteristics on the Earth's surface, including hills, valleys, depressions, steep cliffs, and gentle slopes. Furthermore, the map reader will be capable of identifying regions susceptible to geological risks, like landslides and flooding. Individuals seeking to acquire property, engage in gardening, organize a hiking or camping expedition, or survey land for the construction of a road, dam, or building should initially study a topographic map.

3.2 MAP ORIENTATION AND SCALE

All topographic maps created by the U.S. Geological Survey (U.S.G.S) are orientated with north positioned at the top. Consequently, if you identify a location on the map and proceed towards the upper section, you are advancing in a northerly direction; conversely, if you move towards the lower section, you are heading southward. Movement to the right corresponds to the east, while movement to the left corresponds to the west. These maps are aligned with their edges parallel to lines of longitude, which are imaginary lines encircling the globe and oriented to intersect the northern and southern geographic poles. The 0° longitude line, referred to as the Prime Meridian, traverses Greenwich, England, and extends to 180° in both eastern and western directions (Figure 3.1). Visualizing longitude lines can be facilitated by considering an orange; when peeled, its portions resemble the longitudinal divisions that segment the Earth. All lines of longitude converge at the center of the orange. (or the geographic north and south poles of the Earth).

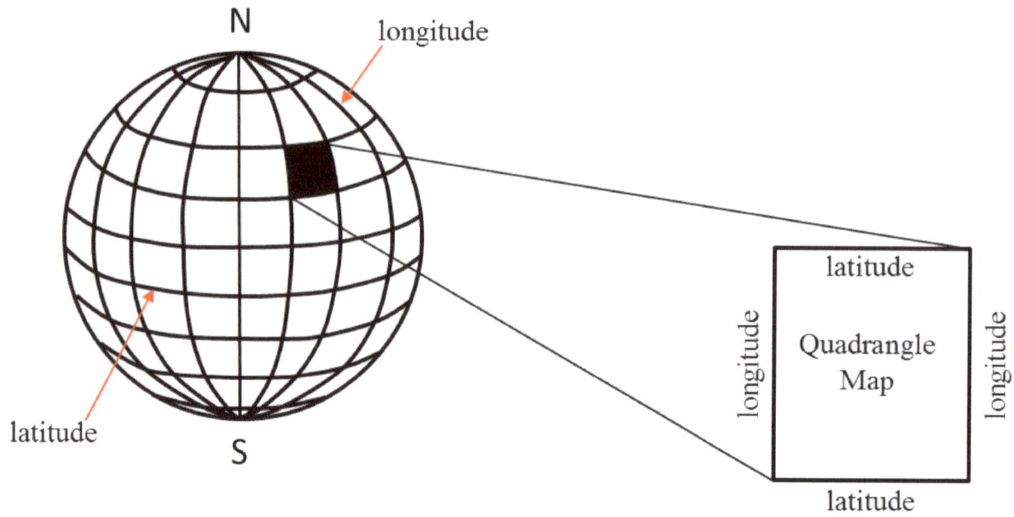

Figure 3.1 | Latitude and Longitude grid system of the Earth.

The upper and lower margins of a topographic map are aligned parallel to lines of latitude, which are fictitious lines encircling the globe and oriented perpendicularly to the Earth's axis. The 0° latitude line is the Equator; latitude lines extend to 90° north or 90° south of the Equator, with the North Pole at 90°N and the South Pole at 90°S.

The grid system of latitude and longitude enables the precise identification of a location on Earth, contingent upon the designation of latitude as either N or S of the Equator, and longitude as E or W of the Prime Meridian. A degree of latitude or longitude signifies a considerable distance on Earth; hence, degrees are subdivided into minutes (note that a minute of distance differs from a minute of time), and these minutes are further divided into seconds. One degree of latitude or longitude comprises 60 minutes (60'), and each minute has 60 seconds (60"). An example of a specific position on the Earth's surface is 33°34'22"N, 85°05'46"W, denoted as "33 degrees, 34 minutes, 22 seconds north latitude, and 85 degrees, 5 minutes, 46 seconds west longitude."

The latitude and longitude coordinates of each topographic map are located at the corners of the map. These maps frequently depict an area of the Earth that is less than one degree in distance. A typical topographic map displays a span of 7.5 minutes (7' 30") for both latitude and longitude; thus, the upper and lower edges of the map denote a distance of 7.5 minutes of latitude, while the left and right edges signify a distance of 7.5 minutes of longitude. Due to the constant degree of distance on the map, a shorthand notation indicating only the variation in minutes and seconds is marked at specific locations along the map's perimeter.

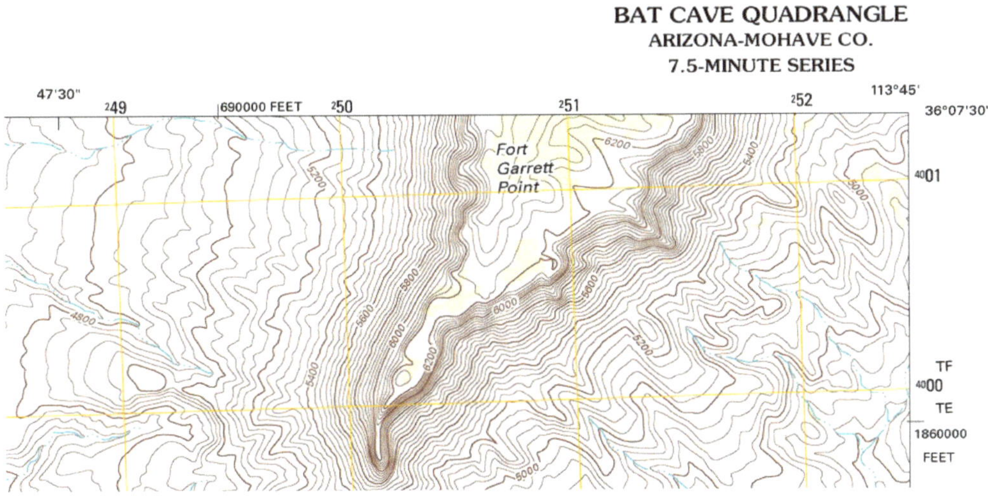

Figure 3.2 | This is the top right corner of a topographic map covering an area of 7' and 30" of latitude and longitude. Note that the coordinates are written out completely in the corner of the map, but shorthand notation is used for the longitude coordinate on the top edge of the map (47'30"). Since the number increased towards the left (by a distance of 2'30"), we know that these are west longitude numbers (meaning the longitude at that tick mark is 113°47'30" W), since only lines of longitude located west of the Prime Meridian increase with

Only at the corners of the map will the degrees be included for the latitude and longitude coordinates. See the example above, which shows the top right corner of a 7.5 minute map or 7.5 minute quadrangle (the shape of most topographic maps). Additional numbers other than latitude and longitude are also shown on the edges of the map; these are a different grid system and will not be explained here.

Remember that the top and bottom edges of a topographic map are oriented parallel to lines of latitude. In Figure 3.2, the top edge of the map is a latitude line, so the coordinates that are changing as you move along the top of the map (in an East-West direction), must be longitude coordinates, since the latitude should not change. The bottom edge of this map is not shown, but you can predict what the latitude on the bottom edge should be: this map is of a region in Arizona, and since Arizona is located north of the Equator the latitude lines should increase from the bottom of the map, towards the top of the map (as all latitude lines increase as you move north of the Equator). For a 7.5-minute map, the increase in latitude from the bottom corner to the top corner should be exactly 0°7'30"N. This gives the bottom corner of the map a latitude of 36°00'00"N.

All maps are a scaled down version of the region of the world that they depict; if this were not the case, then the map that a person must carry would be the exact same size as a city (if it is a city map) or the size of a state (if it is a state map). Imagine trying to carry around with you a map of the entire country! The word "scale" refers to the amount of reduction, and all maps provide a map scale to indicate how much the area on the map has been reduced. Map scales are provided so that a map reader can determine exactly how much distance is actually represented on their map, or to measure the distance between two points on a map, or even to calculate the gradient of a hill or river. The two commonly used map scales on a topographic map are the **bar scale** (or **graphical scale**) and the **fractional scale** (also known as a **ratio scale**).

In Figure 3.3 there are three bar scales; each bar is a graphical representation of distance on the map, and it is up to the map reader to decide if they want to measure distances in kilometers, meters, miles, or feet. To find the distance between any two points on a map, a person could use a piece of paper to transfer the two points down to the bar scale and read the distance directly from the bar scale.

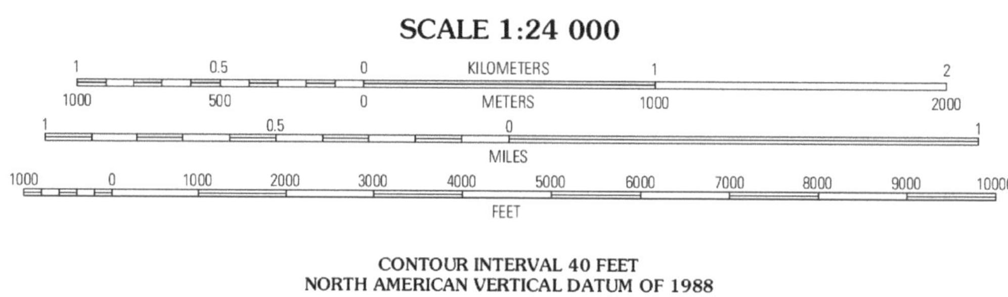

Figure 3.3. Map scales typically located on the bottom of a topographic map. Note that the bar scales start at zero in the interior of each bar for kilometers, miles, and feet. The bar to the left of zero is further subdivided for more accurate distance determinations.

Notice that each bar scale has the starting point (zero) within the interior of the scale, and not on the end of each scale (Figure 3.3).

The other type of map scale is the fractional scale; in the Figure 3.3 the fractional scale is 1:24,000. No units are reported as this ratio of 1 to 24,000 is valid for any unit of measure, provided that it is the same unit. For example, if using inches, then this map scale indicates that 1 inch on the map is actually covering 24,000 inches of ground (the distance between two locations in the real world). Or if using centimeters, then 1 centimeter on the map is actually covering 24,000 centimeters of ground. If our map was the same size as the area that it is representing (say for example, a map of the room you are currently sitting in), then the fractional scale of your map would be 1:1, and your map would be the exact same size as your room! This brings us again to the definition of a map, which is a scaled down version of a region that it is made to represent; maps that are greatly scaled down (greatly reduced) are called small scale maps even though they represent large sections of the Earth. For example, a 1:500,000 map will show a large section of the Earth, but small details are lost (such as building locations or small streets), whereas a 1:12,000 map is a large scale map even though it shows a much smaller region of the Earth's surface, but details can be seen (such as buildings, roads, and other landmarks). Placing your fingertip on the surface of a small scale map may cover an area of several miles, but placing your fingertip on a large scale map (such as 1:12,000) may cover only 1/10 of a mile.

One advantage to using a fractional or ratio scale is that any unit of measure can be used, and conversions are easy to make when needed. For example, for a 1:12,000 map 1 inch on the map is equal to 12,000 inches on the Earth's surface, and since there are 12 inches in 1 foot, we can also say that 1 inch on the map is equal to 1000 feet on the Earth's surface (convert the 12,000 inches to feet by multiplying by the conversion factor 1ft/12in). Simply verbalizing this scale by saying "on this map, 1 inch represents 1000 feet" is a third type of map scale, which for obvious reasons is called a **verbal scale.** Writing the phrase "1 inch equals 1000 feet" is a way of adding a verbal scale to your map.

Table 3.1 | Some useful conversion factors:

1 foot = 12 inches	1 meter = 3.28 feet
1 mile = 5280 feet	1 mile = 63,360 inches
1 kilometer = 1000 meters	1 kilometer = 0.62 miles

Remember that since conversion factors are equalities, such as 1 foot = 12 inches, then dividing one by the other (1ft/12in) gives you 1, and since multiplying anything by 1 does not change any value, all we really are doing is changing the units. Therefore the calculation 5.5 ft x (12in/1foot) will allow 5.5 ft to be expressed as inches, which in this case would be 66 inches.

3.2.1 LAB EXERCISE

Part A – Practice Questions (non-graded)

The following problems are for practice; answers to these questions are provided at the end of the chapter.

1. A 15 minute quadrangle map of a region within the United States with a longitude of 76°00'00" in the right corner of the map, will read what longitude in the left corner?
2. A 15 minute quadrangle map of a region within the United States with a latitude of 43°15'00" in the top corner of the map, will read what latitude in the bottom corner?
3. A fractional scale of 1:24,000 means that 1 inch =____feet.
4. A fractional scale of 1:24,000 means that 1 foot =____kilometers.

3.3 CONTOUR LINES

Contour lines add a vertical dimension to a map, indicating heights above sea level. Each contour line connects places of equal elevation; so, traversing that line in reality indicates maintaining a constant elevation while moving along that hypothetical line. Should you deviate from that line, you are either ascending or descending in elevation. Envision standing on a little circular island in the ocean, ascending from the shoreline to a height of 10 feet above the water's edge. Walking around the island at a consistent altitude of 10 feet above the shore would trace a contour line indicative of 10 feet of elevation above sea level. Deviation from that line indicates either an ascent or a descent. If you ascend an additional 10 feet, maintaining that elevation (now 20 feet above sea level) while traversing the island, you are now following the 20-foot contour line. The vertical change in elevation between these two contiguous contour lines is referred to as the contour interval, which is 10 feet in this instance. Transferring these hypothetical lines onto a map would reveal three lines creating concentric rings that denote 0 ft (the shoreline or sea level), 10 ft, and 20 ft, resulting in a bull's-eye pattern. Congratulations for creating your inaugural topographic map!

A topographic map features contour lines represented in brown, and each map possesses a specific contour interval. The elevations indicated by the contour lines are not consistently identified on every line (see to Figure 3.2). Every fifth contour line is designated with an elevation and is emphasized; this type of contour line is referred to as an index contour. Employing index contours enhances a map's visual appeal, particularly when the contour lines are abundant and tightly spaced.

To ascertain the elevation of each contour line, one must first identify the contour interval of the map. Utilizing the values of two contiguous index contours allows for the straightforward calculation of the contour interval between each line. For instance, there are four contour lines between the 5200ft and 5400ft index contours (refer to Figure 3.2), indicating that these four lines divide the 200ft elevation interval between the index contours into five segments. Dividing the 200 ft elevation change between the index contours by 5 results in a contour interval of 40 ft, analogous to halving a ruler to produce two 6-inch segments or partitioning it into three equal sections to get four 3-inch segments. To confirm this, identify the 5200ft index contour on the western side of the map in Figure 3.2, then increment the elevation by 40ft with each contour line crossed while proceeding eastward towards the 5400ft contour line. Fortunately, it is unnecessary to perform this computation to ascertain the contour interval on a comprehensive topographic map, as all such maps include the contour interval at the bottom near the bar and fractional scales (see to Figure 3.3). The contour interval must be adhered to for each contour line on a map; for instance, if the contour interval is 50 ft, potential contour lines could include 50 ft, 100 ft, 150 ft, 200 ft, and so forth.

You may question the reason for the close spacing of contour lines in certain regions of a map (e.g., the center section of Figure 3.2) and their wider spacing in other areas (e.g., the western section of Figure 3.2). Envision yourself once more on the circular island in the ocean, positioned 10 feet above sea level (at the 10-foot contour line). What distance must you traverse to ascend the hill to attain a 20-foot elevation? The steepness of the hill determines the duration of the ascent; a gradual slope may necessitate a prolonged trek to get an elevation of 20 feet. On a topographic map, the contour lines for this hill would be widely scattered. Conversely, if the hill's gradient is exceedingly high, one need not traverse a considerable distance to attain a 20ft elevation, and the contour lines indicative of such a steep incline will be closely spaced on a topographic map. A slope (gradient) is defined as the change in elevation divided by the distance. You can readily compute the slope of a hill or any area on a topographic map provided you are aware of the elevation change between two places and the distance separating those sites. Gradients are often expressed in feet per mile (ft/mile), however alternative units are also utilized. Utilize the contour lines to ascertain altitudes and the bar scale on your map to gauge distance.

Topographic maps, in addition to contour lines, will provide benchmarks (actual surveyed points) at various locations. These surveyed points are exact elevations above sea level and are widely used to identify the altitudes of mountains, hilltops, road intersections and airport runways. These benchmarks are seldom positioned on a contour line; rather, they are typically marked by a black "x" or denoted with the letters "BM," accompanied by elevations displayed in black numerals (in contrast to the brown numerals on index contours). Benchmark sites are typically situated between contour lines; for instance, a benchmark of 236ft will be located between the 230ft and 240ft contour lines (if the contour interval is 10ft) or between the 235ft and 240ft contour lines (if the contour interval is 5ft).

In addition to obeying the defined contour interval for a map, contour lines should never branch (split) or simply end within of the mapped region. Instead, these lines are continuous, but they may extend beyond the map's boundary. Contour lines also never touch or overlap, unless certain rare cases occur, such as if there is a vertical or overhanging cliff. In the case of a vertical cliff, the contour lines will appear to converge.

The complete third dimension (elevation) depicted by the contour lines on a topographic map is referred to as relief, which may be readily calculated by identifying the greatest and lowest contour line heights and subtracting the two values to ascertain the vertical relief illustrated on the map. The most challenging aspect is locating the highest and lowest elevations on the map. Commence by identifying the highest index contour line and proceed to enumerate the lines until you arrive at the lowest contour line. In Figure 3.2, the uppermost contour line traverses the letter "r" in Fort (of Fort Garrett Point). The identical contour line loops back and intersects the letter "o" in Fort. The elevation of this line is 6360ft (based on the contour interval of 40ft). This represents merely a fraction of a 7.5-minute map (or quadrangle), resulting in the absence of certain identifying elevation numbers on some index contours. Nonetheless, the index contours remain easily identifiable as they are rendered in bold (darkened lines). To identify the lowest elevation on the map, find the lowest index contour line and continue counting lines in the downhill direction. A straightforward method to ascertain the downhill direction is to locate a water feature on the map; water is represented in blue on topographic maps, with moving water, such as rivers or streams, depicted as blue lines. A dashed blue line, as depicted in Figure 3.2, signifies that the stream is dry for a portion of the year, referred to as an intermittent stream. Water accumulates in depressions, such as basins (where ponds, lakes, or seas are situated) or valleys (like stream or river valleys); hence, contour lines should indicate a decline in height as one approaches a water feature on a map. Referring back to Figure 3.2, it is evident that the highest area of the map is the central portion where Fort Garrett Point is located, and that any point west, south and east of this is a downhill direction. All streams are directed away from the Fort Garrett Point area. The minimum elevation will be shown by a contour line intersecting the stream immediately prior to exiting the map region. A meticulous analysis of the contour lines indicates that the lowest contour line is situated in the lower right corner of the image; the contour line intersecting the stream in this area denotes an elevation of 4560 feet.

For the specified segment of the 7.5-minute map depicted in Figure 3.2, the relief of the area is calculated as 6360 ft (highest contour) minus 4560 ft (lowest contour), resulting in a relief of 1800 ft.

A notable characteristic of moving water, such as streams and rivers, is its capacity to erode the landscape, so altering the topography; this is evidenced by the deflection of contour lines on a map when they intersect flowing water. Observe in Figure 3.2 that the contour lines create a "v" shape as they intersect the water, with the pointed end of the "v" oriented upstream. This allows for the straightforward determination of water flow direction, independent of the elevation indicated by the contour lines. Observe in Figure 3.2 that the contour lines intersecting the streams direct towards the center hill (Fort Garrett Point), indicating that the streams are moving away from the central area of the map and towards its periphery.

3.3.1 LAB EXERCISES

Part B Practice Questions

For Questions 5 through 9, refer to Figure 3.4 below, which shows a hill, an intermittent stream, and two index contours (darkened contour lines). Assume the contour interval for this map is 5ft, and the index contour that is crossing the stream has an elevation of 70ft.

5. Which way is the stream flowing, to the North to the South?
6. What is the elevation of the highest contour on this portion of the map?
7. Calculate the relief of this map (Hint: Review the "Contour Lines" section in this chapter for assistance calculating relief).
8. Calculate the gradient of the stream between the highest and lowest contour lines that you can see cross the stream. These two contour lines are 2 miles apart.
9. The hill in the above diagram has a slightly steeper side on which side of the hill, the west or east side?

Figure 3.4 | Portion of the 7.5 minute Quadrangle of Bat Cave, Arizona

You have learned that the spacing between contour lines indicates the slope of the Earth's surface, and that the shape of the contour line as it crosses flowing water can indicate the slope direction. You have also learned that enclosed (circular) contour lines indicate a hill or mountain. However, sometimes there are circular depressions (for example, a sinkhole) found on the Earth's surface and these depressions may appear as hilltops on a topographic map unless a new convention is used. Contour lines with small perpendicular lines (called **hachure marks**) are used for such depressions on a topographic map. The contour interval for the map is still obeyed when contouring a depression. The only difference is that the hachure marks on the contour lines indicate that you should count down in elevation, not up, as you move towards the center of the hachured contour circles. However, if there is a depression at the top of a hill or mountain (for example, a volcanic crater), then the first contour line that is hachured must be the same elevation of the closest contour line that is not hachured. The reason for the repeat is that a person climbing the hill will reach the highest contour line, and walk a little higher still, before descending into the depression (crater), and will therefore encounter the same elevation line while descending (see Figure 3.5).

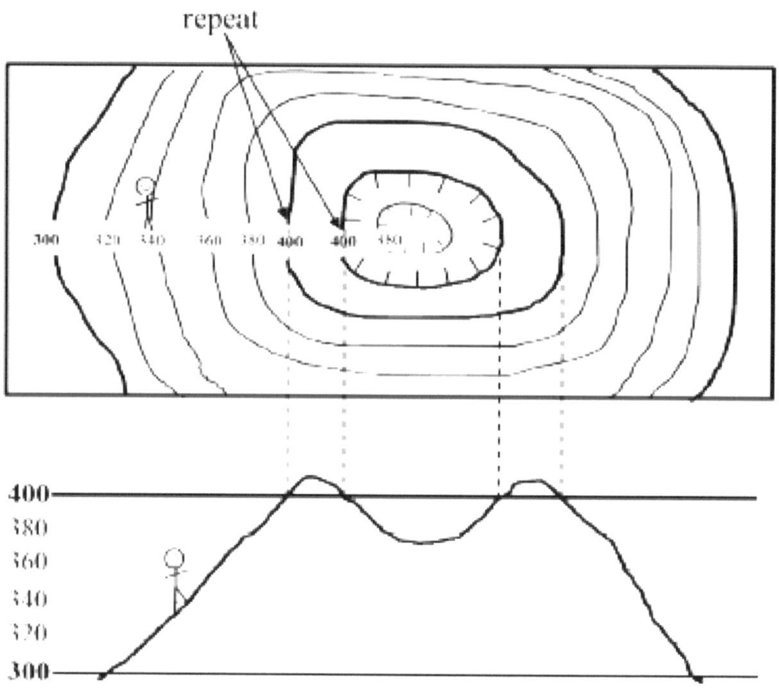

Figure 3.5 | Contours and hachured contours for a depression at the top of a hill. Notice that the first hachured depression is a repeat of the closest non-hachured contour line.

3.4 DEPICTING CONTOUR LINES AND TOPOGRAPHIC PROFILES

Creating a topographic map by delineating contours can be accomplished effortlessly if one adheres to the following principles of contour lines: 1) Contour lines denote connections between points of equivalent elevation above sea level; 2) Contour lines do not intersect, bifurcate, or terminate; 3) Contour intervals must be adhered to, thus contour line elevations can only be multiples of the contour interval; and 4) Contour lines form a "v" pattern when crossing streams and rivers, with the "v" consistently oriented upstream.

When you delineate a contour line on a map, you will observe that the elevations on one side of the line are lower, while those on the opposite side are higher. Upon completing the contour lines, you will observe that certain lines are drawn closer together in specific regions while others are spaced further away, and you may have also delineated an area by creating a contour line in a circular configuration. These circular patterns signify hilltops, as illustrated in the image below (Figure 3.6). To depict the appearance of these hills in profile, or how they would seem from a distance while standing on the ground, one can create a topographic profile. A topographic profile is a lateral depiction of a topographic map, representing only the area indicated by the line on the map. (line A-B on Figure 3.6).

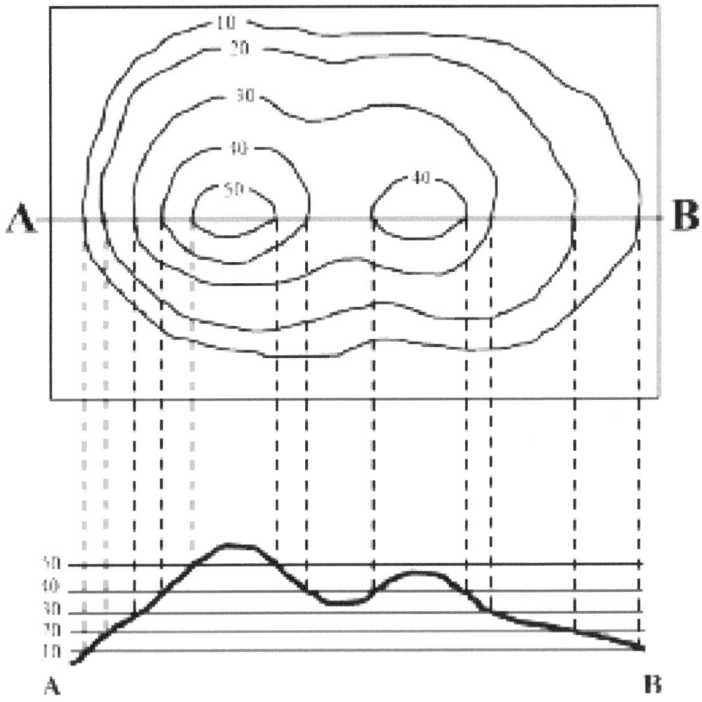

Figure 3.6 | Contour map and topographic profile of two hills and a valley between them.

To construct a profile, you need graph paper, a ruler and a pencil. You want to have the y-axis of the graph paper represent the elevations of the contour lines that intersect your drawn line (line A-B in this case). By using a ruler, you can transfer these elevation points from your topographic map straight down onto your graph paper such as shown in Figure 3.6. Be sure to only plot those elevations that are at the intersection of the contour line with line A-B. Once your points are plotted on the graph paper, you simply connect the dots. As a rule, hill tops will be slightly rounded to show a slight increase in elevation to represent the crest of the hill, but be careful not to draw the hill top too high on your graph paper. For example, the first hill on the left has a top contour line of 50ft. Because there isn't a 60ft contour line on this hill top, we know that the hill's highest point (the crest) is some elevation between 50 and 60ft. When connecting the points on your graph paper in the area between the two hills in the Figure 3.6, you again want to round out the area to represent the base of your valley between the hills, but be careful not to make the valley floor too deep, as according to the topographic map the elevation is below 40ft, but not as low as 30ft.

If you examine the graph showing the topographic profile in Figure 3.6, can you image what would happen to your profile if we changed the spacing for elevations on the y-axis? When the vertical dimension of your graph is different from the horizontal dimension on your map, you may end up with a graph that shows a **vertical exaggeration**, and the features of the Earth represented by your topographic profile may be deformed such as in Figure 3.7.

Sometimes vertical exaggeration is desired, but in some cases you may not want it. To avoid having your profile distorted so that it accurately conveys what the surface of the Earth really looks like in profile, you will want both the vertical and horizontal scales to match. For example, if your map scale is 1 inch = 50 ft, then one inch

on your graph's y-axis should only represent 50ft of elevation. If the topographic map in Figure 3.6 has a fractional scale of 1:12,000 then 1 inch is equal to 12,000 inches or 1000ft; this 1inch = 1000ft equivalency is for the horizontal scale. When we hold a ruler to the y-axis of the topographic profile in Figure 3.6, we see that 0.5 inches = 50ft, which means 1 inch = 100ft on the vertical scale. To calculate the vertical exaggeration in the topographic profile shown in Figure 3.6 we divide the horizontal scale by the vertical scale: (1000ft/1inch)/(100ft/inch) = 10. Therefore the topographic profile in Figure 3.6 represents a profile of the map surface (along the A-B line) that has been vertically exaggerated by 10 times (10X).

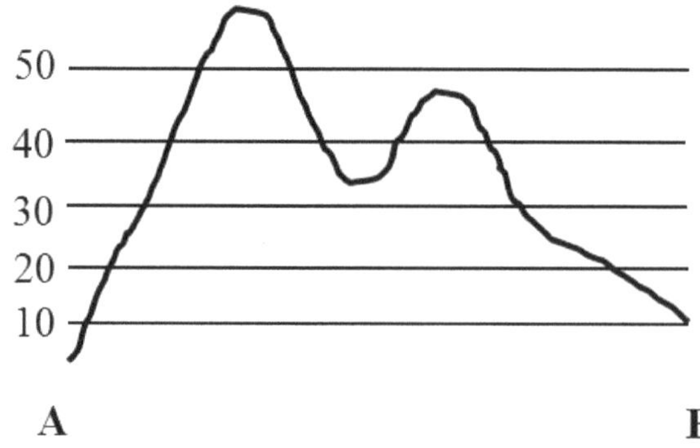

Figure 3.7 | A stretched profile (from Figure 3.6) to demonstrate what vertical exaggeration can do to an image. Notice how much steeper the slopes are in this image.

Answers to Practice Lab Exercises, Parts A and B

1. 76°15'00" W longitude
2. 43°00'00" N latitude
3. 2000 ft
4. 1 ft = 24,000 ft, and 24,000 ft x (1km/3280ft) = 7.32 km
5. South
6. 95 ft (this is the index contour at the top of the hill)
7. 95 ft – 65 ft = 30 ft
8. (80ft – 65 ft)/2miles = 7.5 ft/mile
9. East

3.4.1 LAB EXERCISE

Part C Topographic Maps

This is a graded activity. The following pages must be printed and completed by the student, and mailed to the instructor in order to be scored. Alternatively, you may scan the activity and send it electronically to the instructor; unreadable scans will not be accepted, so be sure to send in legible work. Please remember to include your name in the provided blank.

3.5 TOPOGRAPHIC MAPS LAB ASSIGNMENT

This Lab Assignment is to be mailed to your Instructor at the contact address recorded in the Syllabus. Make sure that you use additional postage if needed. There is no online assessment for the Topographic Maps Lab.

Complete the entire assignment and mail to your instructor postmarked by the assignment deadline. You should make an extra copy to practice on and mail in a clean and neat version for grading. Make sure to include your name on every page and staple all of the pages together.

Please take advantage of all of the resources available to you. Be sure to read the corresponding lecture which contains directions to work out the solutions to the problems below. You should also review the instructional videos located in the unit content area within the course for additional assistance. Finally, check the Topographic Map Unit Discussion forum and the tutor talk area for additional resources and hints.

3.5.1 Topographic Maps Lab

NOTE: For all of the following figures, assume North is up.

1. (10 pts) The following topographic map (Map 3.1) is from a coastal area and features an interesting geological hazard in addition to the Ocean. Using a

contour interval of 40 meters, label the elevation of every contour line on the map below. (Note: elevation is meters above sea level, which makes sea level = _______m).

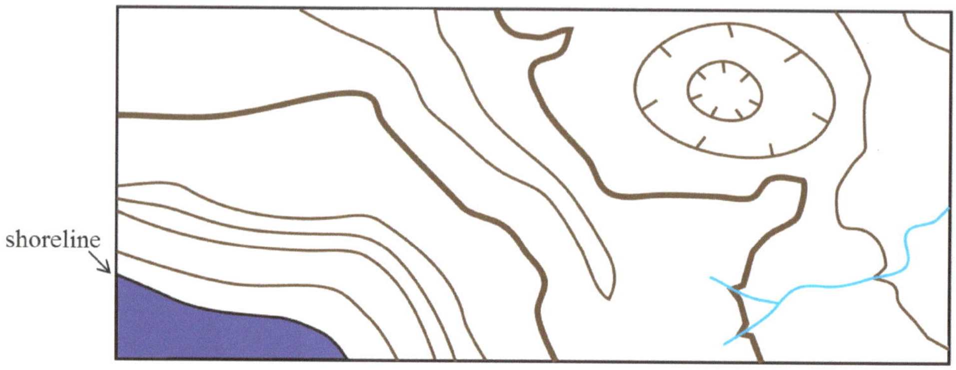

Map 3.1

2. (10 pts) Imagine you are a geologist for the United States Geological Survey. You are tasked with creating your own coastal Topographic map, so you hike around the area with a GPS receiver (Global Positioning System) and every so often you record your position along with the elevation in meters at that point, which results in the following map (Map 3.2). Complete Map 3.2 by adding in the contour lines using a contour interval of 100 meters. Draw the contour lines so that they are continuous (do not die off), and either continue off the map or form an enclosed circle (look at the topographic map in the problem 1 for an example). More often than not, your contour lines will fall between the GPS points on your map, so do your best to determine the contour line positions.

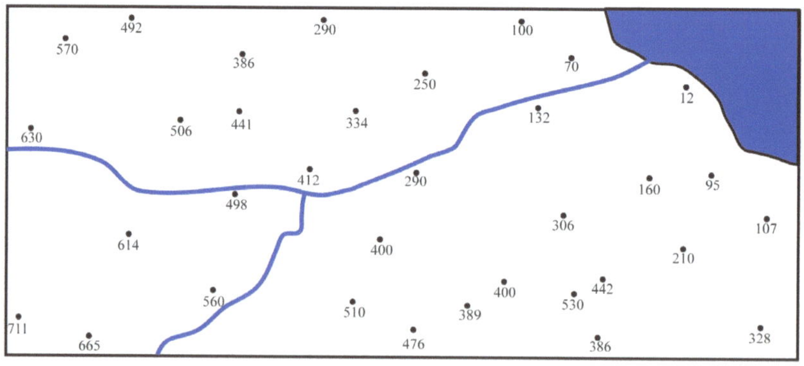

Map 3.2

For questions 3-7 refer to the Map 3.3. The following topographic map shows an interesting and informative geological feature called a drumlin, which is a pile of sediment left behind by a retreating glacier.

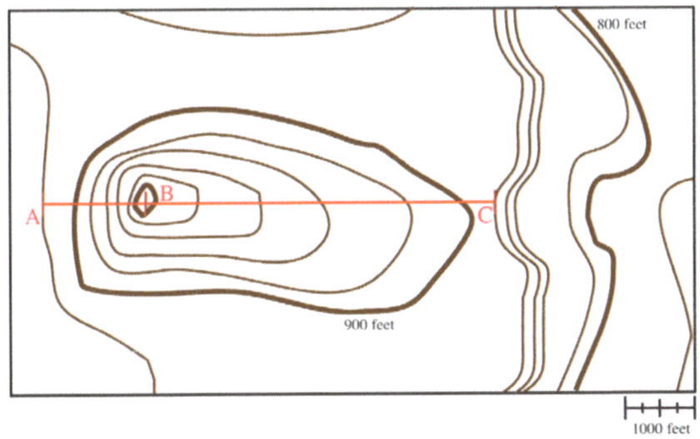

Map 3.3

3. (2 pts) What is the contour interval on Map 3.3?
4. (2 pts) What is the regional relief on Map 3.3?
5. (5 pts) Using the contour lines on Map 3.3, which area along the red line is steeper A to B or B to C? Explain how you came to this conclusion.
6. (5 pts) What is the gradient from A to B and B to C on Map 3.3? Show your work.
7. (2 pts) Drumlins can be used to determine the direction of movement in the glacier with the glacier moving toward the shallower side of the structure. Using your previous answers for Map 3.3, what direction was the glacier traveling? Note: unless indicated otherwise, assume that North is up (towards the top of the map).
8. (20 pts) Construct a topographic profile from A to A' on the graph paper below.

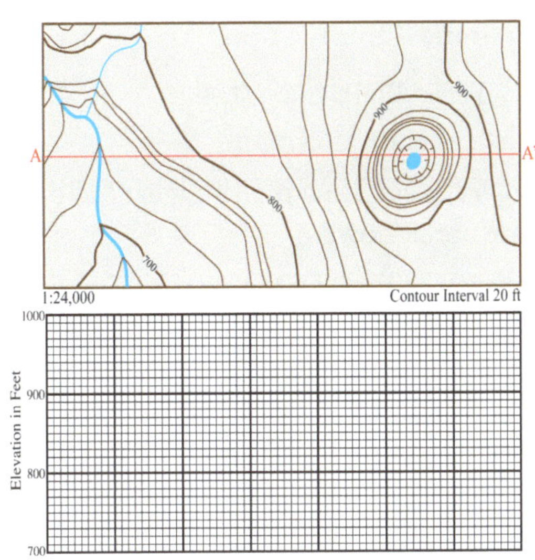

Map 3.4

9. (5 pts) Based on the scale you choose for the topography (vertical axis) in question 8, calculate the amount of vertical exaggeration on the topographic profile you constructed above. Show your work.

For this part of the lab you will need to use Maps 3.5 and 3.6 (appearing at the end of this chapter). Following Maps 3.5 and 3.6 is a Map Key that you can use to identify the various symbols found on topographic maps.Also, note that the maps are in color and the colors have significance in terms of the symbols.

Questions 10-18: Rome North Quadrangle (27 pts)

10. (2 pts) What is the ratio scale of this map?
11. (2 pts) Explain in a sentence how this type of scale works.
12. (2 pts) What is the latitude on the north edge of the map?
13. (2 pts) What is the longitude on the east edge of the map?
14. (6 pts) Find Big Dry Creek, which is north of Rome. What direction does that river flow? Explain two reasons why you came to this conclusion.
15. (2 pts) Examine the large Ridge in the Northwestern portion of the map. What is the tallest point in this ridge? How tall is it?
16. (2 pts) How much higher is that point from Lake Conasauga?
17. (5 pts) What is the gradient between Lake Conasauga and the tallest point in the ridge? Show your work (Hint: zooming out will let you see both features on the map at the same time and may make it easier to measure).
18. (4 pts) How would the gradient change if you measured from Swan Lake to the tallest point in the ridge rather than Lake Conasauga? Explain why.
19. Questions 19-22: Grand Tetons (12 pts)
20. (3 pts) Explain why this map is referred to as a 7.5 minute map?
21. (3 pts) What is the relief on this map?
22. (2 pts) Does Taggert Creek flow into Taggert Lake or Lake Taminah? What direction does the creek flow?
23. (4 pts) Garnet Canyon (a little to the west of the word Garnet) is a common camping location for hikers and mountain climbers at the Grand Teton's National Park. Examine the topography surrounding Garnet Canyon and the Middle Teton. What would be the easiest and safest route from Garnet Canyon to the top of the Middle Teton? Explain why. (Drawing a simple map will help).

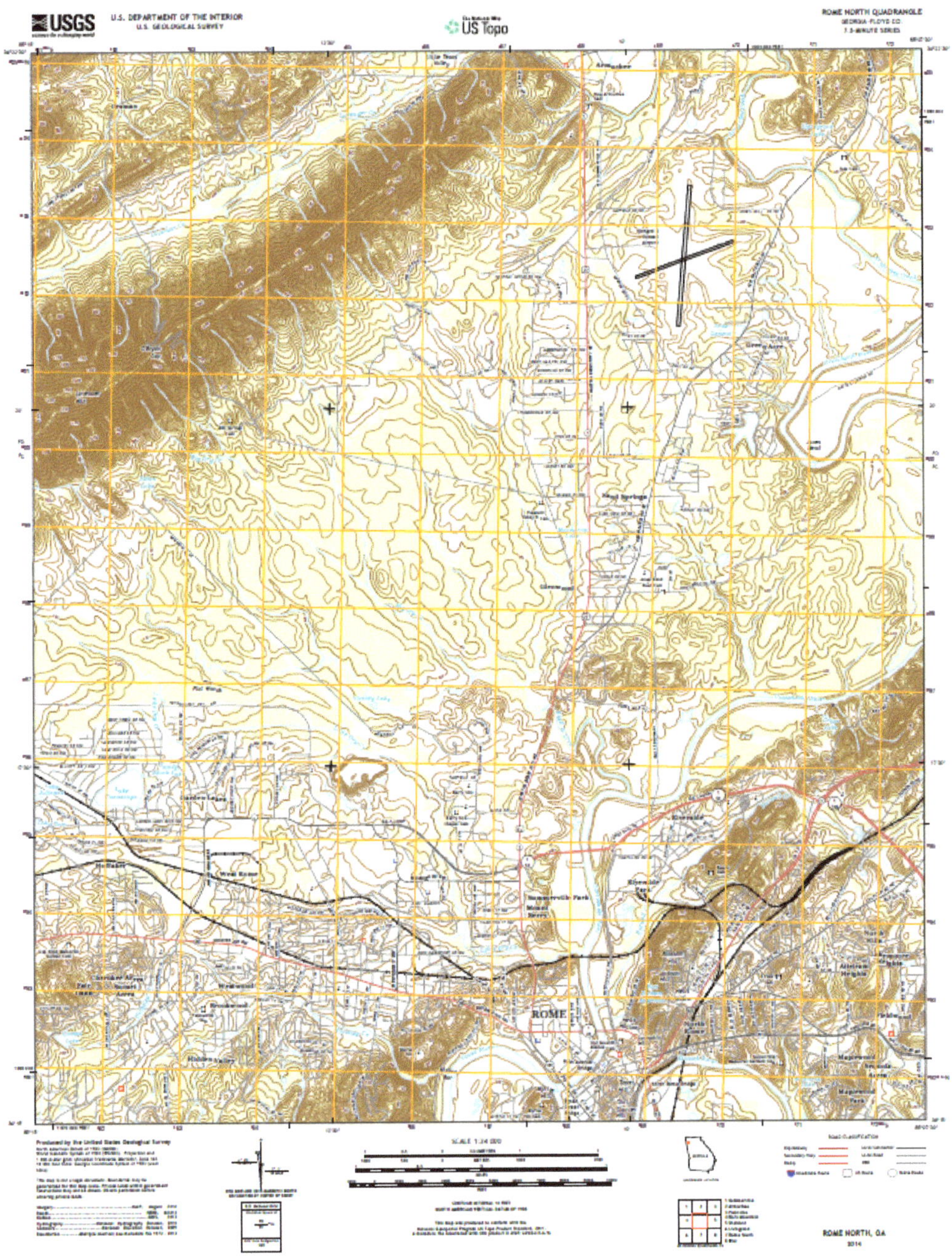

Maps 3.5, 3.6, and Map Key (appearing on following pages)

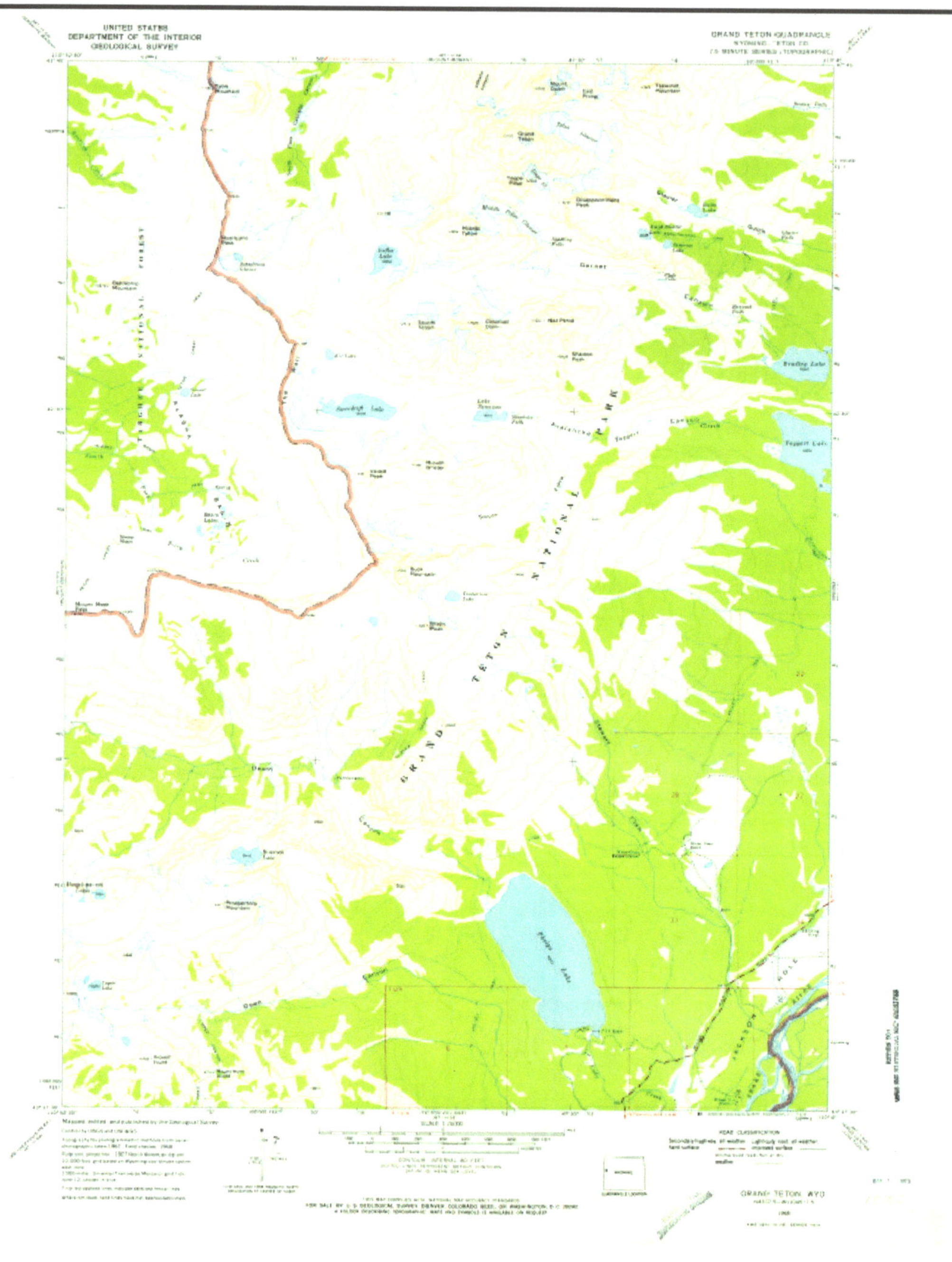
UNITED STATES
DEPARTMENT OF THE INTERIOR
GEOLOGICAL SURVEY
GRAND TETON QUADRANGLE
GRAND TETON NATIONAL PARK
Bradley Lake
Taggart Lake
Phelps Lake

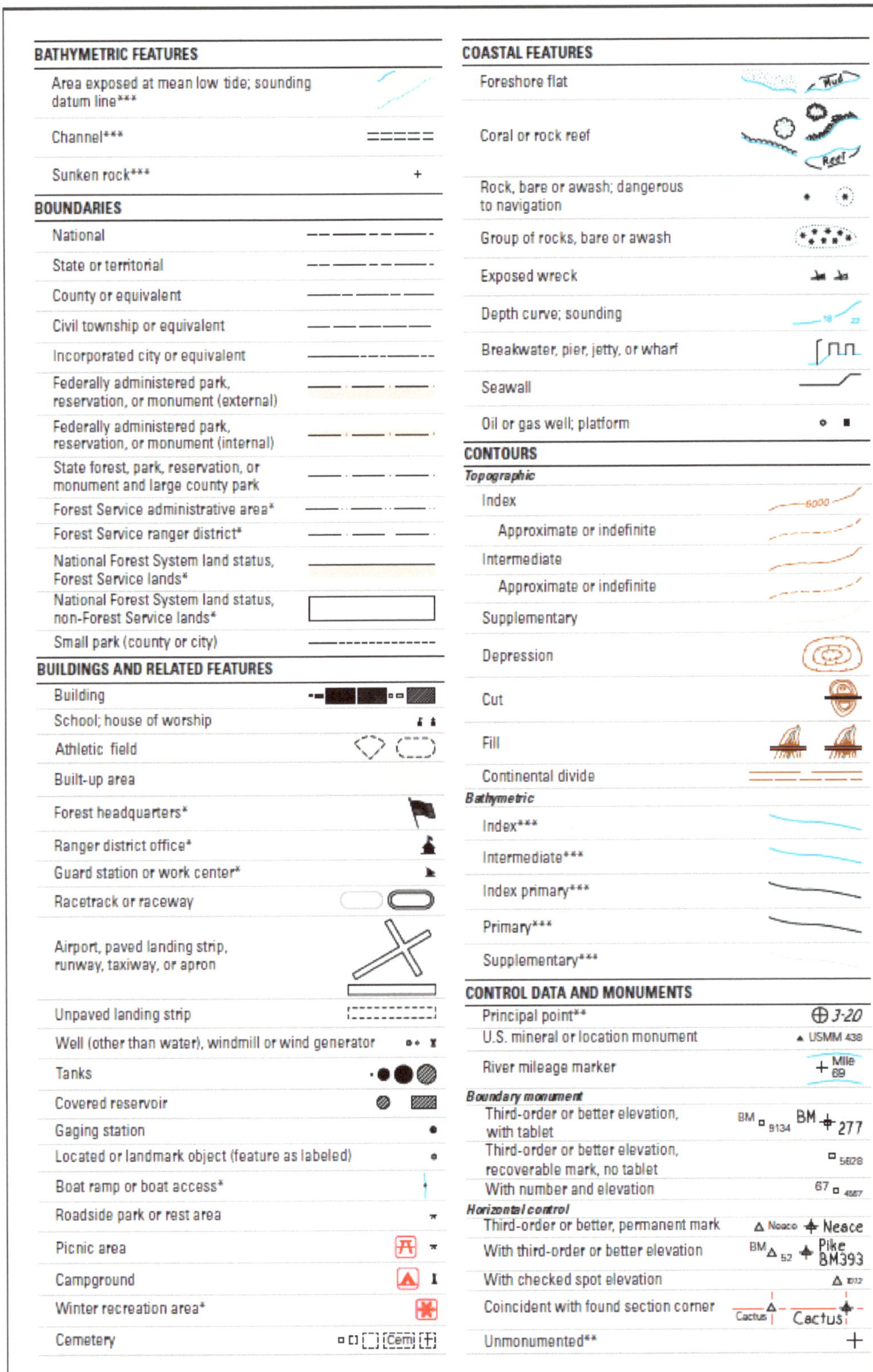
BATHYMETRIC FEATURES
Area exposed at mean low tide; sounding datum line***
Channel***
Sunken rock***
BOUNDARIES
National
State or territorial
County or equivalent
Civil township or equivalent
Incorporated city or equivalent
Federally administered park, reservation, or monument (external)
Federally administered park, reservation, or monument (internal)
State forest, park, reservation, or monument and large county park
Forest Service administrative area*
Forest Service ranger district*
National Forest System land status, Forest Service lands*
National Forest System land status, non-Forest Service lands*
Small park (county or city)
BUILDINGS AND RELATED FEATURES
Building
School; house of worship
Athletic field
Built-up area
Forest headquarters*
Ranger district office*
Guard station or work center*
Racetrack or raceway
Airport, paved landing strip, runway, taxiway, or apron
Unpaved landing strip
Well (other than water), windmill or wind generator
Tanks
Covered reservoir
Gaging station
Located or landmark object (feature as labeled)
Boat ramp or boat access*
Roadside park or rest area
Picnic area
Campground
Winter recreation area*
Cemetery
Cem
COASTAL FEATURES
Foreshore flat
Mud
Coral or rock reef
Reef
Rock, bare or awash; dangerous to navigation
Group of rocks, bare or awash
Exposed wreck
Depth curve; sounding
Breakwater, pier, jetty, or wharf
Seawall
Oil or gas well; platform
CONTOURS
Topographic
Index
6000
Approximate or indefinite
Intermediate
Approximate or indefinite
Supplementary
Depression
Cut
Fill
Continental divide
Bathymetric
Index***
Intermediate***
Index primary***
Primary***
Supplementary***
CONTROL DATA AND MONUMENTS
Principal point**
3-20
U.S. mineral or location monument
USMM 438
River mileage marker
Mile 69
Boundary monument
Third-order or better elevation, with tablet
BM 9134 BM 277
Third-order or better elevation, recoverable mark, no tablet
5628
With number and elevation
67 4567
Horizontal control
Third-order or better, permanent mark
Neace Neace
With third-order or better elevation
BM 52 Pike BM393
With checked spot elevation
Coincident with found section corner
Cactus Cactus
Unmonumented**

CONTROL DATA AND MONUMENTS – *continued*

Vertical control

Third-order or better elevation, with tablet	BM × 5280
Third-order or better elevation, recoverable mark, no tablet	× 528
Bench mark coincident with found section corner	BM 5280
Spot elevation	× 7523

GLACIERS AND PERMANENT SNOWFIELDS

Contours and limits
Formlines
Glacial advance
Glacial retreat

LAND SURVEYS

Public land survey system

Range or Township line	
Location approximate	
Location doubtful	
Protracted	
Protracted (AK 1:63,360-scale)	
Range or Township labels	R1E T2N
Section line	
Location approximate	
Location doubtful	
Protracted	
Protracted (AK 1:63,360-scale)	
Section numbers	1 - 36
Found section corner	
Found closing corner	
Witness corner	WC
Meander corner	MC
Weak corner*	

Other land surveys

Range or Township line
Section line
Land grant, mining claim, donation land claim, or tract
Land grant, homestead, mineral, or other special survey monument
Fence or field lines

MARINE SHORELINES

Shoreline
Apparent (edge of vegetation)***
Indefinite or unsurveyed

MINES AND CAVES

Quarry or open pit mine	
Gravel, sand, clay, or borrow pit	
Mine tunnel or cave entrance	
Mine shaft	
Prospect	
Tailings	Tailings
Mine dump	
Former disposal site or mine	

PROJECTION AND GRIDS

Neatline	39°15′ 90°37′30″
Graticule tick	55′
Graticule intersection	
Datum shift tick	

State plane coordinate systems

Primary zone tick	640 000 FEET
Secondary zone tick	247 500 METERS
Tertiary zone tick	260 000 FEET
Quaternary zone tick	98 500 METERS
Quintary zone tick	320 000 FEET

Universal transverse metcator grid

UTM grid (full grid)	273
UTM grid ticks*	269

RAILROADS AND RELATED FEATURES

Standard guage railroad, single track
Standard guage railroad, multiple track
Narrow guage railroad, single track
Narrow guage railroad, multiple track
Railroad siding
Railroad in highway
Railroad in road
Railroad in light duty road*
Railroad underpass; overpass
Railroad bridge; drawbridge
Railroad tunnel
Railroad yard
Railroad turntable; roundhouse

RIVERS, LAKES, AND CANALS

Perennial stream
Perennial river
Intermittent stream
Intermittent river
Disappearing stream
Falls, small
Falls, large
Rapids, small
Rapids, large
Masonry dam
Dam with lock
Dam carrying road

RIVERS, LAKES, AND CANALS – *continued*

Feature	Symbol
Perennial lake/pond	
Intermittent lake/pond	
Dry lake/pond	Dry Lake
Narrow wash	
Wide wash	Wash
Canal, flume, or aqueduct with lock	
Elevated aqueduct, flume, or conduit	
Aqueduct tunnel	
Water well, geyser, fumarole, or mud pot	
Spring or seep	

ROADS AND RELATED FEATURES

Please note: Roads on Provisional-edition maps are not classified as primary, secondary, or light duty. These roads are all classified as improved roads and are symbolized the same as light duty roads.

Feature	Symbol
Primary highway	
Secondary highway	
Light duty road Light duty road, paved* Light duty road, gravel* Light duty road, dirt* Light duty road, unspecified*	
Unimproved road Unimproved road* 4WD road 4WD road* Trail	
Highway or road with median strip	
Highway or road under construction	Under Const
Highway or road underpass; overpass	
Highway or road bridge; drawbridge	
Highway or road tunnel	
Road block, berm, or barrier*	
Gate on road*	
Trailhead*	

SUBMERGED AREAS AND BOGS

Feature	Symbol
Marsh or swamp	
Submerged marsh or swamp	
Wooded marsh or swamp	
Submerged wooded marsh or swamp	
Land subject to inundation	Max Pool 431

SURFACE FEATURES

Feature	Symbol
Levee	Levee
Sand or mud	Sand
Disturbed surface	
Gravel beach or glacial moraine	Gravel
Tailings pond	Tailings Pond

TRANSMISSION LINES AND PIPELINES

Feature	Symbol
Power transmission line; pole; tower	
Telephone line	Telephone
Aboveground pipeline	
Underground pipeline	Pipeline

VEGETATION

Feature	Symbol
Woodland	
Shrubland	
Orchard	
Vineyard	
Mangrove	Mangrove

4 Chapter PLATE TECTONICS

4.1 INTRODUCTION: AN OVERVIEW

In chapter one, we examined the scientific process and defined a theory as a well-substantiated explanation for a natural occurrence that remains unproven in its entirety. A Grand Unifying Theory constitutes a fundamental framework within a discipline, exemplified by the theory of gravity in physics or the theory of evolution in biology. The Grand Unifying Theory of geology is Plate Tectonics, which characterizes the Earth's outer layer as a brittle shell divided into mobile segments known as tectonic plates. This idea is substantiated by numerous lines of evidence, including the configuration of continents, the distribution of fossils and rocks, the presence of environmental indicators, and the locations of mountains, volcanoes, trenches, and seismic activity. The movement of tectonic plates is observable within human timescales and may be accurately detected with GPS satellites.

Plate tectonics is essential to geology as it facilitates the reconstruction of Earth's past. This idea elucidates the formation of the initial continents, the genesis of seas, and contributes to hypotheses regarding the origin of life. The idea elucidates the geographic distribution of geological phenomena, including mountains, volcanoes, rift valleys, and trenches. Ultimately, it aids in evaluating the potential hazards of geological disasters, including earthquakes and volcanoes, worldwide. The strength of this theory resides in its capacity to formulate testable hypotheses about Earth's history and to make predictions concerning its future.

4.2 PROOF OF CONTINENTAL DRIFT

The notion that the continents seem to have been contiguous due to their shapes is not novel; it was first articulated in the writings of Sir Francis Bacon in 1620. The derived hypothesis from this observation is clear: the continents' forms interlock because they were formerly contiguous and have subsequently separated and migrated. This theory pertains to a historical occurrence and cannot be empirically tested without a time machine. Consequently, geoscientists restructured the idea by positing that the continents were once interconnected and inquired about further patterns we may anticipate discovering. This is precisely how early 20th-century Earth scientists, including Alfred Wegener, approached this significant scientific inquiry.

Wegener aggregated rock types, fossil occurrences, and environmental indicators from the geological record across many continents, particularly Africa and South America, which seemingly were once contiguous, revealing striking parallels. Subsequent scientists contributed, enabling the scientific community to assemble a comprehensive dataset demonstrating that the continents were once interconnected in a supercontinent known as Pangaea (a term introduced by Alfred Wegener) and had gradually migrated to their present locations over time. The dating of these rocks, utilizing the methods outlined in chapter one, enabled scientists to get insights into the rate of motion, so aiding in the elucidation of the principles that govern plate tectonics.

4.2.1 LAB EXERCISE

This lab will use two different ways to input your answers. Most of the questions will be multiple choice and submitted online as you have in previous labs. Other questions will give you a blank box to input your answer as text. Your professor will manually grade this text, such that the format is not as important as your answer. This format allows you the opportunity to show your work using simple symbols and allows the instructor to better see your thought process. Also note, that for many of these questions there is not a single correct answer and seeing your thought process and understanding the material is more important than your answer. Therefore, it is important to show your work.

In addition, several of the exercises that follow use Google Earth. For each question (or set of questions) paste the location that is given into the "search" box. Examine each location at multiple eye altitudes and differing amounts of tilt. For any measurements use the ruler tool, this can be accessed by clicking on the ruler icon above the image.

Part A – Plate Motion and Evidence

As was mentioned above one of the most striking things about the geography of the continents today is how they appear to fit together like puzzle pieces. The reason for this is clear: they once were connected in the past and have since separated shifting into their current positions. Open Google Earth and zoom out to an eye altitude of ~8000 miles. Examine the coastlines of eastern South America and Western Africa and notice how well they match in shape.

There are scientifically important rock deposits in southern Brazil, South America and Angola, Africa that show the northernmost glacial deposits on the ancient continent of Pangaea, which indicates these two areas were once connected. Based on the shape of the two coastlines, give the present day latitude and longitude of two sites along the coast of these countries that used to be connected when the two continents were joined as a part of Pangaea (note: there are multiple correct answers):

1. Brazil (Latitude and Longitude)
2. Angola (Latitude and Longitude)
3. Measure in centimeters the distance (Map Length) between the two points you recorded in the previous question. Given that this portion of Pangaea broke apart 200,000,000 years ago, calculate how fast South America and Africa are separating in cm/year? (Hint: Speed= Distance/Time)
4. When will the next supercontinent form? Examine the Western Coast of South America, the Eastern Coast of Asia, and the Pacific Ocean. If South America and Africa are separating and the Atlantic Ocean is growing, then the opposite must be occurring on the other side of the earth (the Americas are getting closer to Asia and the Pacific Ocean is shrinking). How far apart are North America and Mainland Asia in cm? (measure the distance across the Pacific at 40 degrees north latitude basically measure between Northern California and North Korea)? Take that distance and divide it by the speed you calculated in question 3 to estimate when the next supercontinent will form. Show your work!

Use the Figures 4.1 and 4.2 to answer questions 5-7.

Figure 4.1

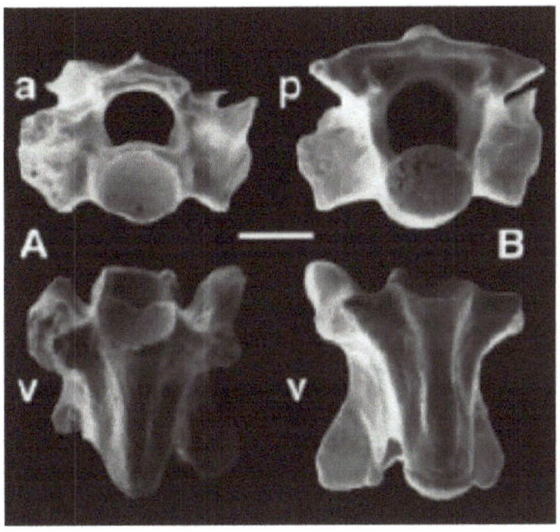

Figure 4.2

Figures 4.1 and 4.2 | The distribution across Australia and Antartica (Figure 4.1) of the fossil snake *Patagoniophis* (Figure 4.2). Obviously, this small snake was unable to swim the immense distance between the contients and, therefore, lived while Australia and Antarctica were still joined together. Figure modification from the Australia Department of Natural Resources and Scanlon (2005), Memoirs of the Queensland Museum.

5. How far have the snake fossils moved apart since they were originally deposited?

   a. 1250 miles b. 1700 miles c. 2150 miles d. 2700 miles

6. Given that this portion of the Australian plate moves at a speed of 2.2 inches per year, how old are the snake fossils?

   a. 310 million years old
   b. 217 million years old
   c. 98 million years old
   d. 62 million years old
   e. 34 million years old

7. There are fossils such as *Glossopteris* and *Lystrosaurus* that are found in rocks in South America and Africa that indicate they were part of Pangaea approximately 200 million years ago. These same fossils can be found in Australia, which indicates it, along with Antarctica, was also part of Pangaea at that time. Based on your answer to question 6 which of the following statements about the break-up of Pangaea is TRUE?

    a.Australia and Antarctica separated before the break-up of Pangaea.

    b.Australia and Antarctica separated during the break-up of Pangaea.

    c.Australia and Antarctica separated after the break-up of Pangaea.

4.3 HOT SPOTS

A further line of evidence for monitoring plate motion is the positioning of hot spots. Hot spots are volcanically active regions on the Earth's surface resulting from unusually heated mantle rocks beneath. This heat originates from a mantle plume ascending from the depths of the mantle, leading to the melting of rocks and the formation of volcanoes. These mantle plumes originate deep within the Earth, rendering them impervious to the movements of continents or the oceanic crust. Mantle plumes seem to remain fixed over time; nevertheless, as the tectonic plate shifts over the hotspot, a succession of volcanoes is generated. This provides geologists with a comprehensive perspective on the movement of a tectonic plate throughout time, as the distribution of volcanoes indicates the direction of motion, while their ages elucidate the rate of movement.

A notable example of a hot spot is located beneath Hawaii. The mantle plume produces heat that leads to an active volcano on the crust's surface. Each eruption contributes to the volcano's growth until it ultimately breaches the ocean's surface, resulting in the formation of an island. As the tectonic plate travels away from the hotspot, the volcano loses its thermal energy and becomes dormant. The volcano then cools, contracts, erodes, gradually submerges beneath the water surface, and is transported by the tectonic plate over time. As each island migrates away from the mantle plume, a new island will subsequently be generated at the hot site, creating a continuous conveyor belt of islands. The scars of ancient islands in Hawaii provide a remarkable perspective on the movement of the tectonic plate beneath the Pacific Ocean.

4.3.1 LAB EXERCISE

Materials

Type "Hawaii" into the search bar of Google Earth and examine the chain of Hawaiian Islands. On a separate sheet of paper please draw yourself a map of the islands and label the following on your map (making sure to include the names), which will be used to answer the following questions.

Islands to include:

Big Island, Maui, Kauai, Nihoa (23 03 32.79N 161 55 11.94W)

a. Put a North arrow on your map.
b. Label the ages of each of the islands determined by radiometric dating of the lava.

Big Island0 (active), Maui – 1.1 million, Kauai4.7 million, Nihoa7.2 million years

a. Place a dot at the center of each island and measure on Google earth the distance between the center of an island and its adjacent island in centimeters (for example measure the distance between the center of the Big Island and Maui).
b. Look closely at each island in Google Earth and record their maximum elevation in centimeters. Remember elevation can be found by placing your cursor over a point and reading the elevation on the lower right of the image by the latitude and longitude. The elevation will be given in meters, but can be converted to centimeters by multiplying by 100. (Hint: tilting the image of the island will help to find the highest point.)

Part B Hawaii

8. Consider the ages and positions of the islands listed above along with what you know about plate tectonics and hotspots. In what general direction is the Pacific Plate moving?

   a. Northwest b. Southeast c. Northeast d. Southwest

9. How fast was the Pacific plate moving during the last 1.1 million years between the formation of the Big Island and Maui in cm/year? To calculate this divide the distance (in centimeters) between the two islands by the difference in their ages.

10. How fast was the Pacific plate moving from 7.2 million years ago to 4.7 million years ago between the formation of Kauai and Nihoa in cm/year? To calculate this divide the distance (in centimeters) between the two islands by the difference in their ages.

11. Examine the headings of the measurements that you took for the previous two questions. The headings indicate the direction the Pacific Plate is moving over the hot spot. How does the direction of motion of the Pacific Plate during the last 1.1 million years differ from direction of movement between 4.7 and 7.2 million years ago? The direction of plate movement in the last 1.1 million years______.

    a shows no change

    b. has become more southerly

    c. has become more northerly

12. Zoom out and examine the dozens of sunken volcanoes out past Nihoa, named the Emperor Seamounts. As one of these volcanic islands on the Pacifi Plate moves off the hotspot it becomes inactive, or extinct, and the island begins to sink as it and the surrounding tectonic plate cool down. The speed the islands are sinking can be estimated by measuring the difference in elevation between two islands and dividing by the difference in their ages (this method assumes the islands were a similar size when they were active). Calculate how fast the Hawaiian Islands are sinking, by using the ages and elevations of Maui and Nihoa.

13. Using the speed you calculated in the previous question (and ignoring possible changes in sea level), when will the Big Island of Hawaii sink below the surface of the ocean? Divide the current maximum elevation of the Big Island by the rate you calculated in the previous question.

14. Now zoom out to ~4000 miles eye altitude and look at the chain of Hawaiian Islands again. Notice the chain continues for thousands of miles up to Aleutian Islands (between Alaska and Siberia). Examine the northernmost sunken volcano (50 49 16.99N 167 16 36.12E) in this chain. Where was that volcano located when it was still active, erupting, and above the surface of the ocean?

    a. 50 49 16.99N 167 16 36.12E
    b 52 31 48.72N 166 25 43.14W
    c. 27 45 49.27N 177 10 08.75W
    d. 19 28 15.23N 155 19 14.43W

4.4 PLATE MATERIALS

Currently, numerous lines of evidence indicate the movement of tectonic plates, along with various supplementary evidences. To formulate a theory, some explanation or process is required to elucidate the observed patterns. The plate tectonics theory posits that the Earth's exterior rigid layer, known as the lithosphere, is fragmented into segments referred to as tectonic plates, which move autonomously over the semi-fluid mantle beneath (Asthenosphere).

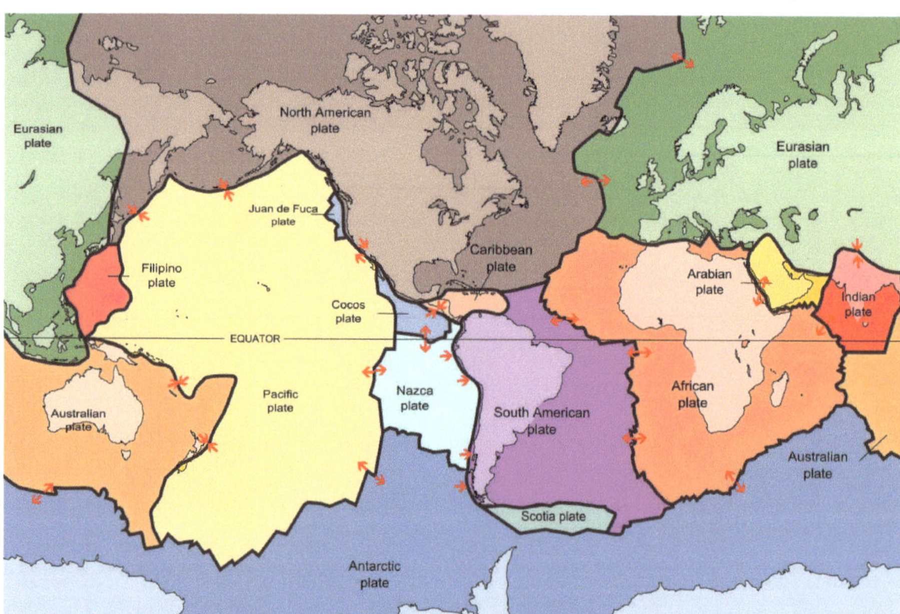

Figure 4.3 | Tectonic plates on Earth.

Tectonic plates consist of the crust and the uppermost mantle, functioning as a brittle solid. These plates may consist of oceanic crust, continental crust, or a combination of both. The Oceanic Crust is thinner and often underlies the world's oceans, whereas the Continental Crust is thicker and, as its name suggests, comprises the continents. The interaction of tectonic plates underlies several geological events and phenomena, necessitating an understanding of their structure to comprehend their interactions more effectively. The interaction of these plates is governed by the relative motion of the two plates (whether converging, diverging, or sliding past each other) and the composition of the crustal segment of the plate (continental or oceanic crust). The continental crust predominantly resembles the composition of granite, an igneous rock characterized by its solid, silica-rich crystalline structure, typically comprising a blend of pink (feldspar), milky white (feldspar), clear (quartz), and black (biotite) crystals. The oceanic crust predominantly consists of the igneous rock gabbro, a solid crystalline rock rich in iron and magnesium, made of a blend of black and dark gray minerals, specifically pyroxene and feldspar.

The variation in rock composition leads to unique physical properties that you will assess in the forthcoming questions..

4.4.1 LAB EXERCISE

Part C – Plate Densities

An important property of geological plates is their density (mass/volume). Remember the asthenosphere has fluid-like properties, such that tectonic plates 'float' relative to their density. This property is called isostasy and is similar to buoyancy in water. For example, if a cargo ship has a full load of goods it will appear lower than if it were empty because the density of the ship is on average higher. Therefore, the relative density of two plates can control how they interact at a boundary and the types of geological features found along the border between the two plates. Measuring the density of rocks is fairly easy and can be done by first weighing the rocks and then calculating their volume. The latter is best done by a method called fluid displacement using a graduated cylinder. Water is added to the cylinder and the level is recorded, a rock is then added to the cylinder and the difference in water levels equals the volume of the rock. Density is then calculated as the mass divided by the volume (Figure 4.4).

The information needed to calculate density was collected for four rocks and can be used to answer the following questions including the weight (in grams) as well as the volume of water recorded by a graduated cylinder (in milliliters) before and after the rock was added. Note: each line on the graduated cylinder represents 10 ml. When measuring the volume please round to the nearest 10 milliliter line on the graduated cylinder. **Hint:** Surface tension will often cause the water level to curve up near the edges of the graduated cylinder creating a feature called a meniscus. To accurately measure the volume, use the lowest point the water looks to occupy.

Figure 4.4 | Method to find the density of a rock. First the weight is measured on a digital scale and then the fluid displacement method is used to determine the volume.

Figure 4.5 | Figure to use to answer questions 15-19.The first row shows images of the four rocks. The second and third rows show the volume (in milliliters) of material in the graduated cylinder before and after the rock was added. The last row shows the mass (in grams) of the four rocks.

15. The rock that most closely resembles the composition of continental crust based on the description in the previous section is:

    a. A   b. B   c. C   d. D

16. Based on the choice you made for question 15, what is the density of the rocks that make up continental crust? Please give your answer in grams/milliliter.

17. The rock that most closely resembles the composition of oceanic crust based on the description in the previous section is:

    a. A b. B c. C d. D

18. Based on the choice you made for question 17, what is the density of the rocks that make up oceanic crust? Please give your answer in grams/milliliter.

19. Remember, because of isostasy the denser plate will be lower than the less dense plate. If oceanic and continental crust collided, based on their densities the_crust would sink below the__crust.

    a. continental; oceanic b. oceanic; continental

4.5 PLATE BOUNDARIES

Tectonic plates can interact in three distinct manners: they may converge, diverge, or laterally glide past one another (Figure 4.6). Another significant aspect is the composition of the interacting plates (oceanic or continental crust), as discussed in the preceding section. The three types of motions, combined with the plate types on either side of the boundary, can generate significantly diverse structures and geological events (Table 4.1).

Plates that are separating from one another are termed Divergent. Divergent borders are significant as they facilitate the separation of continents into distinct plates and the formation of fresh oceanic crust. When a divergent boundary develops within a continent, that region undergoes extensional deformation. This leads to the area being less substantial, resulting in a topographic depression or valley. The extension is not a seamless process, rendering the area susceptible to earthquakes and volcanic activity. Ultimately, the crust becomes sufficiently thin to rupture, creating a fissure between the tectonic plates, which will be filled with molten rock, resulting in the formation of new oceanic crust. A slender and dense plate will be topographically depressed and will be inundated, creating a long and narrow sea. As the tectonic plates continue to diverge, new crust is perpetually generated at the plate boundary along an elevated ridge termed a mid-oceanic ridge.

Two plates that are moving toward each other are termed Convergent. Convergent borders are significant since they facilitate the suturing of separate tectonic plates to create bigger plates and serve as sites for the destruction of oceanic crust. The features observed at convergent frontiers are contingent upon the types of tectonic plates involved. When two thick, low-density continental plates meet, a significant impact occurs, resulting in the formation of mountains. This is a tumultuous process leading to numerous earthquakes, rock deformation (folds and faults), and the elevation of mountains. The rocks are subjected to significant pressure and heat, ultimately fusing into a singular plate. When a continental plate and an oceanic plate converge at a continent-ocean convergent plate boundary, subduction occurs, causing the oceanic plate to descend beneath the continental plate. This will produce multiple features, including a profound trench adjacent to the subducting plate, numerous earthquakes, and the generation of magma, leading to a chain of volcanoes along the coastline. The Wadati-Benioff zone, linked to this sort of plate boundary, is a region where earthquakes occur, extending in depth from shallow (near the trench) to deep (~600 km), signifying the subduction of the oceanic plate into the mantle. The convergence of two oceanic plates will lead to subduction, exhibiting analogous characteristics to those previously mentioned. The sole exception is that the volcanoes will manifest on an oceanic plate and ultimately create islands along the tectonic border.

The sliding of two plates past one another is referred to as a Transform boundary. This boundary is distinct from the preceding two as it neither generates new crust nor obliterates old crust. Consequently, there will be less prominent geological features. Transform boundaries are frequently characterized by numerous earthquakes, which may occur near the surface, and by specific river patterns that become displaced due to the underlying ground movement. Transform boundaries are frequently linked to mid-oceanic ridges. If the ridge possesses a jagged or stair-stepped edge, the divergence of the two tectonic plates will also induce transform motion, as illustrated in Figure 4.6.

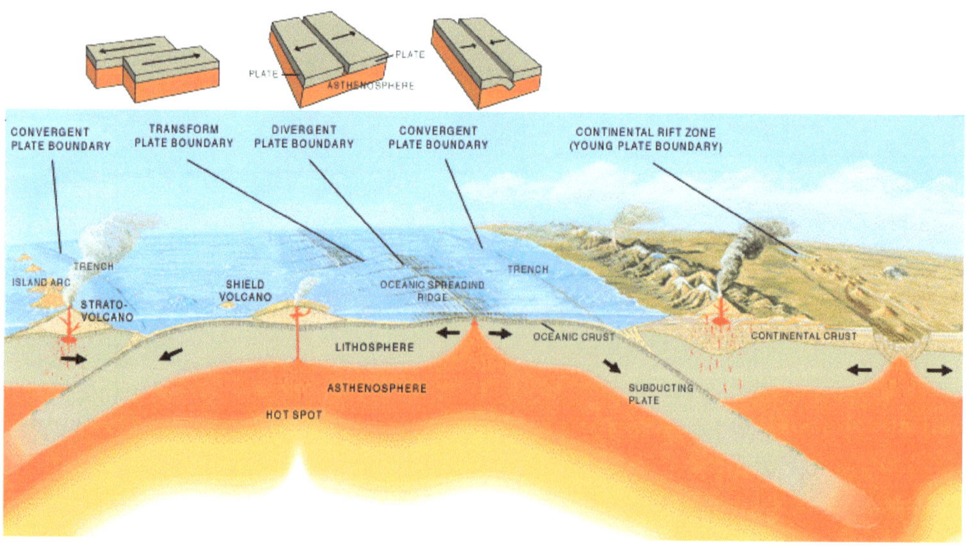

Figure 4.6 | Figure showing multiple plate boundaries and the features associated with them.

Table 4.1. Characteristics of the different Plate Boundaries

Boundary Type	Plate Compositions	Earthquake Depth	Change in Crust	Identifying Features
Divergent				
	Continent-Continent	Shallow	No change	Rift Valley and Volcanoes
	Ocean-Ocean	Shallow	Formation of New Crust	Ocean Ridges
Convergent				
	Continent-			
	Continent Continent-	Shallow to Inter-mediate	Metamorphism of Crust	Mountains
	Ocean	Shallow to Deep	Melting of Crust	Trench and Coastal Volcanoes
	Ocean-Ocean	Shallow To Deep	Melting of Crust	Trench and Volcan-ic Islands
Transform				
	Continent-Continent	Shallow	No change	Offset rivers
	Ocean-Ocean	Shallow	No change	Often associated with Ocean ridges.

4.5.1 LAB EXERCISE

Magma is formed from the melting of rock at both convergent and divergent boundaries. However, the processes that occur to melt the rock are quite different. Three different processes are involved in the melting of rocks as we will explore in the following exercise. In Figure 4.7 you can see a graph depicting a variety of temperature and pressure conditions. The increasing temperature with pressure on rocks as you go deeper within the earth through the crust and mantle lithosphere is called the geothermal gradient (shown in black). This gradient shows the actual temperature conditions that exist in the lithosphere. Obviously, the addition or subtraction of heat or pressure can move rocks off that gradient and cause potential change. The orange line represents the temperature and pressure required for a dry mantle rock to start to melt and any point to the right of this line is where melting of lithospheric rock can occur. The blue line represents the temperature and pressure required for a lithospheric rock to melt if water is present.

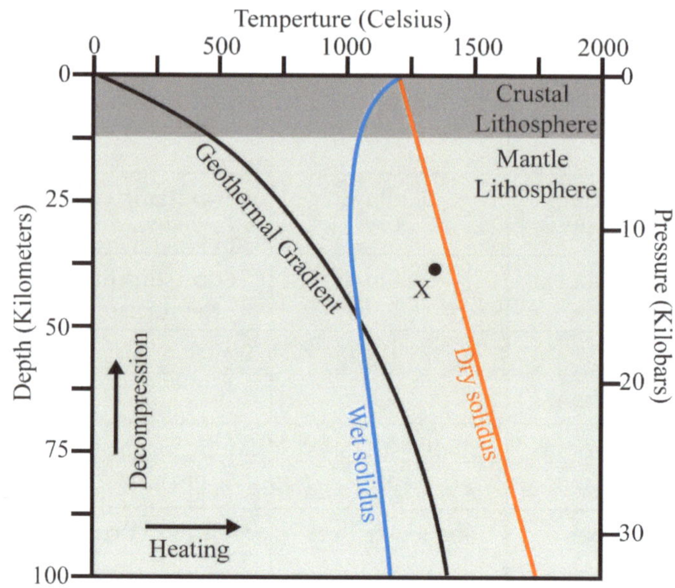

Figure 4.7 | Melting diagram for mantle rock.

20. According to the geothermal gradient, rocks buried 75 km beneath the surface would normally be at what temperature?

At 75 km depth, rocks will be heated to about ______degrees Celsius.

a. 1500 b. 1250 c. 1000 d. 750

21. According to the geothermal gradient, rocks at 500 degrees Celsius will be buried how deep?

At 500 degrees Celsius, rocks will be buried to about _______km depth.

a. 8 b. 12.5 c. 20 d. 27

22. What is the physical state of a dry mantle rock at point X?

a. Completely melted b. Starting to melt

c. Completely solid

23. What happens when the lithosphere at point X is heated to 1500 °C?

a. No change b. Starts to crystallize c. Starts to melt

24. At what depth will the dry mantle rock at point X begin to melt if it is uplifted closer to Earth's surface and its temperature remains the same?

a. 35 km b. 25 km c. 18 km d. 12 km

25. What would happen to the mantle rock at point X if water is added to it?

a. No change b. Starts to crystallize c. Starts to melt

Part E – Boundaries

Earthquakes are great indicators of plate boundaries and are associated with all three boundary types. One type of boundary is unique in having a Wadati-Benioff zone. Answer the following questions using Figure 4.8.

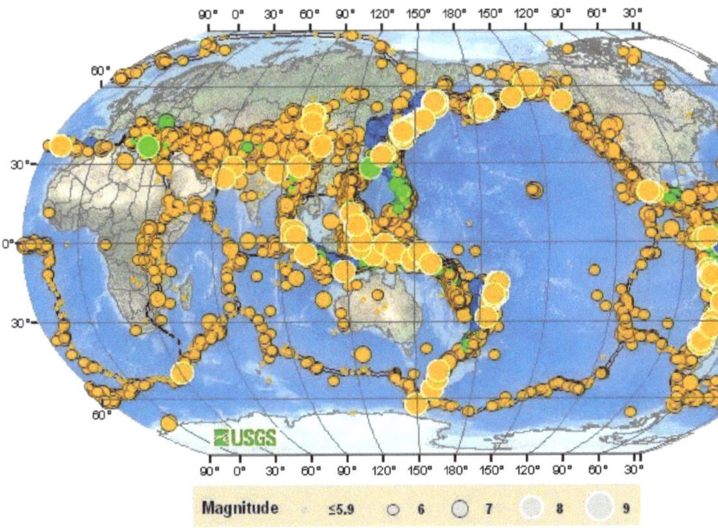

Figure 4.8 | Distribution and depth of Earthquakes.

26. Which of the following places represent a Wadati-Benioff zone?
    a. 10°S, 110°W b. 0°, 0° c. 15°S, 180° d. 30°N, 75°E

27. The Wadati-Benioff zone is associated with which type of plate boundary?
    a.Divergent
    b.Convergent (Continent-Continent)
    c.Convergent (Continent-Ocean or Ocean-Ocean)
    d.Transform

Type 34 46 16.16 N 118 44 58.19 W into the search bar in Google Earth and zoom out to an eye altitude of 10 miles. Quail Lake is a dammed river that is sitting directly over top of the San Andres Fault, which is a well-known transform boundary with the North American Plate on the northern side and the Pacific Plate on the southern side. This boundary is running East-West in this area and you may be able to see the boundary better by zooming out.

28. Examine the path of the river that feeds into and flows out of Quail Lake. What direction is the North American plate moving in comparison to the Pacific Plate at this location?
    a. East b. West

29. Given that San Francisco is located on the North American Plate and Los Angles is located on the Pacific Plate, are these two cities getting closer together or farther apart over time?

    a. Closer   b. Farther

Google Earth: Identifying Plate Boundaries

Re-read the section on Plate Boundaries before answering the following questions.

30. Type "15 19 48.78 S 75 12 03.41 W" into the Google Earth Search bar. What type of tectonic plates are present?

    a. Ocean Ocean   b. Ocean Continent

    c. Continent Continent

31. What type of plate tectonic boundary is present?

    a. Transform   b. Convergent   c. Divergent

32. Type "6 21 49.68 S 29 35 37.87 E" into the Google Earth Search bar. What type of process is going on at this location?

    a. Seafloor spreading   b. Continental rifting

    c. Subduction

33. What features would you expect to occur at this type of boundary?

    a. Earthquakes and a trench   b. Volcanoes and a valley

    c. Mountains and landslides   d. Earthquakes and offset rivers

34. Type "28 04 27.04N 86 55 26.84E" into the Google Earth Search bar. What type of tectonic plates are present?

    a. Ocean Ocean   b. Ocean Continent

    c. Continent Continent

35. What type of plate tectonic boundary is present?

    a. Transform   b. Convergent   c. Divergent

36. Type "46 55 25.66 N 152 01 25.17 E" into the Google Earth Search bar. What type of tectonic plates are present? Make sure to zoom out to get a good view of the relevant features.

    a. Ocean Ocean   b. Ocean Continent

    c. Continent Continent

37. What features would you expect to occur at this type of boundary?

    a. Volcanos, earthquakes and a trench

    b. Volcanoes and a linear valley

    c. Mountains and landslides

    d. Earthquakes and offset rivers

38. Type "43 41 07.81 N 128 16 56.29 W" into the Google Earth Search bar. What type of tectonic plates are present? Hintmake sure to re-read the section on plate boundaries before answering!

    a. OceanOcean b. OceanContinent c. ContinentContinent

39. What type of plate tectonic boundary is at this exact location?

    a. Transform b. Convergent c. Divergent

40. This plate boundary isn't as simple as the previous examples, meaning another nearby plate boundary directly influences it. Zoom out and examine the area, what other type of boundary is nearby?

    a. Transform b. Convergent c. Divergent

4.6 PLATE TECTONIC MECHANISMS

The question still remains, why do tectonic plates move? The answer comes down to gravity and mantle convection. You have already studied in chapter two how the mantle flows through time creating convection currents. These convection currents flow underneath the plates and through friction pull them along at the surface as well as when they are subducted which is a force called **slab suction**. Related to this force is **slab pull,** which is a gravitational force pulling the cold subducting plate down into the mantle at a subduction zone. In addition, there is a force from potential energy at ocean ridges called **ridge push**. This is a gravitational force pushing down on the elevated ridge and because of the plates curvature it results in a horizontal force pushing the plate along the earth's surface. These forces all occur deep inside the Earth and operate on very large geographic scales making them difficult to measure. There are several competing models for the mechanisms behind plate motion, such that there are still some areas of debate surrounding the mechanics of plate tectonics which is why Plate Tectonics is a scientific theory. Documenting an event is much easier and more straightforward than explaining why it occurred.

4.6.1 LAB EXERCISE

Part F – Mechanisms

41. Go back to the location in Google Earth that you examined for question 36 (46 55 25.66 N 152 01 25.17 E). Which of the three proposed plate tectonic mechanisms would NOT occur at this location?

    a. Slab pull b. Ridge push c. Slab suction

4.7 STUDENT RESPONSES

The following is a summary of the questions in this lab for ease in submitting answers online.

1. Brazil (Latitude and Longitude)

2. Angola (Latitude and Longitude)
3. Measure in centimeters the distance (Map Length) between the two points you recorded in the previous question. Given that this portion of Pangaea broke apart 200,000,000 years ago, calculate how fast South America and Africa are separating in cm/year? (Hint: Speed= Distance/Time)
4. When will the next supercontinent form? Examine the Western Coast of South America, the Eastern Coast of Asia, and the Pacific Ocean. If South America and Africa are separating and the Atlantic Ocean is growing, then the opposite must be occurring on the other side of the earth (the Americas are getting closer to Asia and the Pacific Ocean is shrinking). How far apart are North America and Mainland Asia in cm? (measure the distance across the Pacific at 40 degrees north latitudebasically measure between Northern California and North Korea)? Take that distance and divide it by the speed you calculated in question 3 to estimate when the next supercontinent will form. Show your work!
5. How far have the snake fossils moved apart since they were originally deposited?

   a. 1250 miles b. 1700 miles c. 2150 miles d. 2700 miles
6. Given that this portion of the Australian plate moves at a speed of 2.2 inches per year, how old are the snake fossils?

   a. 310 million years old b. 217 million years old

   c. 98 million years old d. 62 million years old

   e. 34 million years old
7. There are fossils such as *Glossopteris* and *Lystrosaurus* that are found in rocks in South America and Africa that indicate they were part of Pangaea approximately 200 million years ago. These same fossils can be found in Australia, which indicates it, along with Antarctica, was also part of Pangaea at that time. Based on your answer to question 6 which of the following statements about the break-up of Pangaea is TRUE?

   a.Australia and Antarctica separated before the break-up of Pangaea.

   b.Australia and Antarctica separated during the break-up of Pangaea.

   c.Australia and Antarctica separated after the break-up of Pangaea.
8. Consider the ages and positions of the islands listed above along with what you know about plate tectonics and hotspots. In what general direction is the Pacific Plate moving?

   a. Northwest b. Southeast c. Northeast d. Southwest
9. How fast was the Pacific plate moving during the last 1.1 million years between the formation of the Big Island and Maui in cm/year? To calculate this divide the distance (in centimeters) between the two islands by the difference in their ages.
10. How fast was the Pacific plate moving from 7.2 million years ago to 4.7 million years ago between the formation of Kauai and Nihoa in cm/year? To calculate this divide the distance (in centimeters) between the two islands by the difference in their ages.

11. Examine the headings of the measurements that you took for the previous two questions. The headings indicate the direction the Pacific Plate is moving over the hot spot. How does the direction of motion of the Pacific Plate during the last 1.1 million years differ from direction of movement between 4.7 and 7.2 million years ago? The direction of plate movement in the last 1.1 million years______.

    a. shows no change

    b. has become more southerly

    c. has become more northerly

12. Zoom out and examine the dozens of sunken volcanoes out past Nihoa, named the Emperor Seamounts. As one of these volcanic islands on the Pacific Plate moves off the hotspot it becomes inactive, or extinct, and the island begins to sink as it and the surrounding tectonic plate cool down. The speed the islands are sinking can be estimated by measuring the difference in elevation between two islands and dividing by the difference in their ages (this method assumes the islands were a similar size when they were active). Calculate how fast the Hawaiian Islands are sinking, by using the ages and elevations of Maui and Nihoa.

13. Using the speed you calculated in the previous question (and ignoring possible changes in sea level), when will the Big Island of Hawaii sink below the surface of the ocean? Divide the current maximum elevation of the Big Island by the rate you calculated in the previous question.

14. Now zoom out to ~4000 miles eye altitude and look at the chain of Hawaiian Islands again. Notice the chain continues for thousands of miles up to Aleutian Islands (between Alaska and Siberia). Examine the northernmost sunken volcano (50 49 16.99N 167 16 36.12E) in this chain. Where was that volcano located when it was still active, erupting, and above the surface of the ocean?

    a. 50 49 16.99N 167 16 36.12E

    b 52 31 48.72N 166 25 43.14W

    c. 27 45 49.27N 177 10 08.75W

    d. 19 28 15.23N 155 19 14.43W

15. The rock that most closely resembles the composition of continental crust based on the description in the previous section is:

    a. A b. B c. C d. D

16. Based on the choice you made for question 15, what is the density of the rocks that make up continental crust? Please give your answer in grams/milliliter.

17. The rock that most closely resembles the composition of oceanic crust based on the description in the previous section is:

    a. A b. B c. C d. D

18. Based on the choice you made for question 17, what is the density of the rocks that make up oceanic crust? Please give your answer in grams/milliliter.

19. Remember, because of isostasy the denser plate will be lower than the less dense plate. If oceanic and continental crust collided, based on their densities the_crust would sink below the__crust.

    a. continental; oceanic   b. oceanic; continental

20. According to the geothermal gradient, rocks buried 75 km beneath the surface would normally be at what temperature?

At 75 km depth, rocks will be heated to about _degrees Celsius.

    a. 1500   b. 1250   c. 1000   d. 750

21. According to the geothermal gradient, rocks at 500 degrees Celsius will be buried how deep?

At 500 degrees Celsius, rocks will be buried to about _km depth.

    a. 8   b. 12.5   c. 20   d. 27

22. What is the physical state of a dry mantle rock at point X?

    a.Completely melted   b. Starting to melt

    c.Completely solid

23. What happens when the lithosphere at point X is heated to 1500 °C?

    a. No change   b. Starts to crystallize   c. Starts to melt

24. At what depth will the dry mantle rock at point X begin to melt if it is uplifted closer to Earth's surface and its temperature remains the same?

    a. 35 km   b. 25 km   c. 18 km   d. 12 km

25. What would happen to the mantle rock at point X if water is added to it?

    a. No change   b. Starts to crystallize   c. Starts to melt

26. Which of the following places represent a Wadati-Benioff zone?

    a. 10°S, 110°W   b. 0°, 0°   c. 15°S, 180°   d. 30°N, 75°E

27. The Wadati-Benioff zone is associated with which type of plate boundary?

    a.Divergent

    b.Convergent (Continent-Continent)

    c..Convergent (Continent-Ocean or Ocean-Ocean)

    d.Transform

28. Examine the path of the river that feeds into and flows out of Quail Lake. What direction is the North American plate moving in comparison to the Pacific Plate at this location?

    a. East   b. West

29. Given that San Francisco is located on the North American Plate and Los Angles is located on the Pacific Plate, are these two cities getting closer together or farther apart over time?

a. Closer b. Farther

30. Type "15 19 48.78 S 75 12 03.41 W" into the Google Earth Search bar. What type of tectonic plates are present?

a. OceanOcean b. OceanContinent c. ContinentContinent

31. What type of plate tectonic boundary is present?

a. Transform b. Convergent c. Divergent

32. Type "6 21 49.68 S 29 35 37.87 E" into the Google Earth Search bar. What type of process is going on at this location?

a. Seafloor spreading b. Continental rifting

c. Subduction

33. What features would you expect to occur at this type of boundary?

a. Earthquakes and a trench b. Volcanoes and a valley

c. Mountains and landslides d. Earthquakes and offset rivers

34. Type "28 04 27.04N 86 55 26.84E" into the Google Earth Search bar. What type of tectonic plates are present?

a. OceanOcean b. OceanContinent c. ContinentContinent

35. What type of plate tectonic boundary is present?

a. Transform b. Convergent c. Divergent

36. Type "46 55 25.66 N 152 01 25.17 E" into the Google Earth Search bar. What type of tectonic plates are present? Make sure to zoom out to get a good view of the relevant features.

a. OceanOcean b. OceanContinent c. ContinentContinent

37. What features would you expect to occur at this type of boundary?

a. Volcanos, earthquakes and a trench

b. Volcanoes and a linear valley

c. Mountains and landslides

d. Earthquakes and offset rivers

38. Type "43 41 07.81 N 128 16 56.29 W" into the Google Earth Search bar. What type of tectonic plates are present? Hintmake sure to re-read the section on plate boundaries before answering!

a. OceanOcean b. OceanContinent c. ContinentContinent

39. What type of plate tectonic boundary is at this exact location?
    a. Transform b. Convergent c. Divergent
40. This plate boundary isn't as simple as the previous examples, meaning another nearby plate boundary directly influences it. Zoom out and examine the area, what other type of boundary is nearby?
    a. Transform b. Convergent c. Divergent
41. Go back to the location in Google Earth that you examined for question 36 (46 55 25.66 N 152 01 25.17 E). Which of the three proposed plate tectonic mechanisms would NOT occur at this location?
    a. Slab pull b. Ridge push c. Slab suction

REFERENCE

Scalon, J. D. 2005. Australia's oldest known snakes: *Patagoniophis*, *Alamitophis* and *c.f. Madtsoia* (Squamata: Madtsoiidae) from the Eocene of Queensland. Memoirs of the Queensland Museum 51: 215-235.

Chapter 5 WATER

5.1 INTRODUCTION: AN OVERVIEW

Consider the several occasions each day when you take water for granted - you presume the tap will dispense water when you activate the faucet, you anticipate precipitation to irrigate your lawn, and you rely on water for your leisure activities. Water is essential for numerous biological processes and serves as a significant geological agent. Water may significantly alter the environment over time by both eroding canyons and depositing substantial layers of debris. Certain processes are gradual and lead to landscapes eroded over time. Other phenomena, such as floods, can be both dramatic and perilous.

What occurs to water during a rainstorm? Envision yourself in a parking lot adjacent to verdant expanses. What is the destination of the water from the parking lot? A significant portion will flow off as sheet flow and ultimately converge with a stream. What occurs to the precipitation in the grassy region? A significant portion will permeate or absorb into the soil. This lab will address both surface water and groundwater. Both are essential components of the hydrological cycle, wherein water is perpetually recycled through the atmosphere, terrestrial surfaces, and subsequently returned to the oceans. This solar-driven cycle functions efficiently as water can rapidly transition from liquid to gas (or water vapor) under surface conditions.

Surface and groundwater are advantageous for potable water, industry, agriculture, recreation, and trade. The need for water will escalate with population growth, necessitating the protection of both surface and subterranean water supplies.

5.2 STREAMFLOW AND COMPONENTS OF A STREAM

The flowing water in a stream will erode and transport material inside its course, including dissolved chemicals resulting from chemical weathering. Solid sediments can vary in size from microscopic clay and silt particles imperceptible to the naked eye to larger sand and gravel-sized sediments. Even boulders have been transported by large rivers. The little particles maintained in suspension by the flow of water are referred to as suspended load. Larger particles generally move as bed load, rolling along the streambed (Figure 5.1). Dissolved, suspended, and bed loads may traverse extensive distances (e.g., from the headwaters of the

Mississippi River in Minnesota to the Gulf of Mexico at New Orleans); however, they will ultimately settle or deposit. Stream-deposited sediments, known as alluvium, can be deposited at any time, although they predominantly occur during flood episodes. A stream requires energy to carry sediment more efficiently. This energy mostly depends on the volume of water and its velocity, as a rapid stream may transport greater quantities and larger particles of material. As a stream diminishes its energy and decelerates, sediment will be deposited.

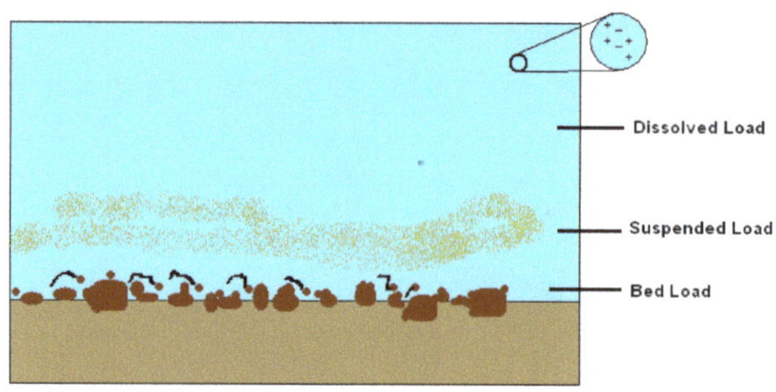

Figure 5.1 | An illustration depicting dissolved, suspended, and bed load.

Under typical circumstances, water will persist within a stream channel. When the water level in a stream surpasses its banks, the overflow will experience a rapid drop in velocity due to increased friction. As its velocity decreases, it will deposit the bigger sandy sediments it is transporting along the channel banks, forming ridges of sandy alluvium known as natural levees (Figure 5.2). As multiple flooding events transpire, these ridges accumulate with successive deposition. These levees constitute a component of a more extensive geomorphological feature referred to as a floodplain. A floodplain is the comparatively level terrain bordering a stream that is prone to inundation during periods of elevated flow. (Figure 5.2).

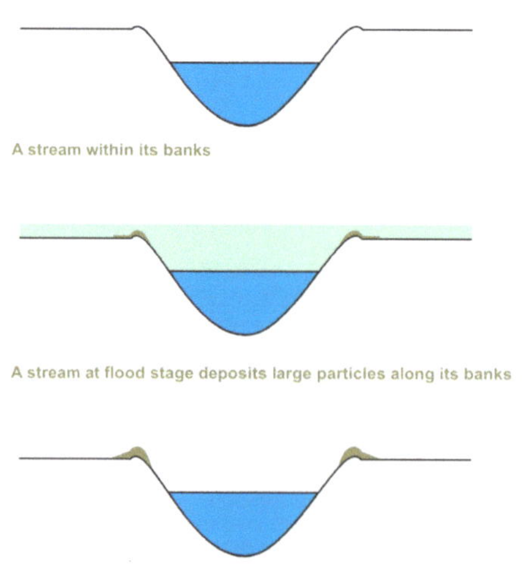

Figure 5.2 | After many floods, natural levees have built up along stream banks.

5.2.1 Stream Drainage Basins and Configurations

The drainage basin of a stream encompasses all terrain that is drained by that stream, including its tributaries, which are the minor streams that contribute to the main stream. You are currently situated in a drainage basin. Are you aware of which one it is? Information is accessible online. Visit the Environmental Protection Agency's website (epa.gov) and search for "Surf Your Watershed" to obtain information. The elevated regions that delineate drainage basins are referred to as drainage divides. The continental divide in the Rocky Mountains delineates the flow of water in North America, directing drainage westward to the Pacific Ocean and eastward to the Gulf of Mexico.

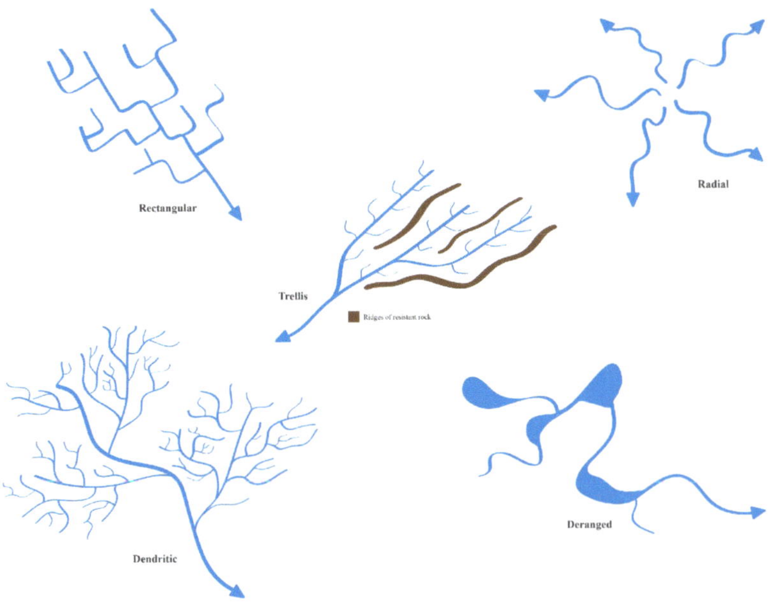

Figure 5.3 | Drainage patterns.

As water traverses rock, it is affected by it. Water seeks to flow towards areas of less resistance, therefore gravitating towards softer rock rather than harder, more resistant rock. This may lead to distinctive drainage patterns. Common drainage patterns include:

- **Dendritic** – this drainage pattern indicates uniformly resistant bedrock that often includes horizontal rocks. Since all the rock is uniform, the water is not attracted to any one area, and spreads out in a branching pattern, similar to the branches of a tree.
- **Trellis** – this drainage pattern indicates alternating resistant and nonresistant bedrock that has been deformed (folded) into parallel ridges and valleys. The water is attracted to the softer rock, and appears much like a rose climbing on a trellis in a garden.

- **Radial** – this drainage pattern forms as streams flow away from a central high point, such as a volcano, resembling the spokes in a wheel.
- **Rectangular** – this drainage pattern forms in areas in which rock has been fractured or faulted which created weakened rock. Streams are then attracted to the less resistant rock and create a network of channels that make right-angle bends as they intersect these breaking points. This pattern will often look like rectangles or squares.
- **Deranged** – this drainage pattern does not follow the rules. It consists of a random pattern of stream channels characterized by irregularity. It indicates that the drainage developed recently and has not had time to form one of the other drainage patterns yet.

5.3 STREAM GRADIENT AND THE CYCLE OF STREAM EROSION

Stream gradient denotes the incline of the stream's course, calculated as rise over run. It is the vertical descent of the stream across a horizontal expanse. You have already encountered gradients in topographic maps. The calculation can be performed using the subsequent equation.:

Gradient = (change in elevation) / distance

Let's calculate the gradient from A to B in Figure 5.4 below.

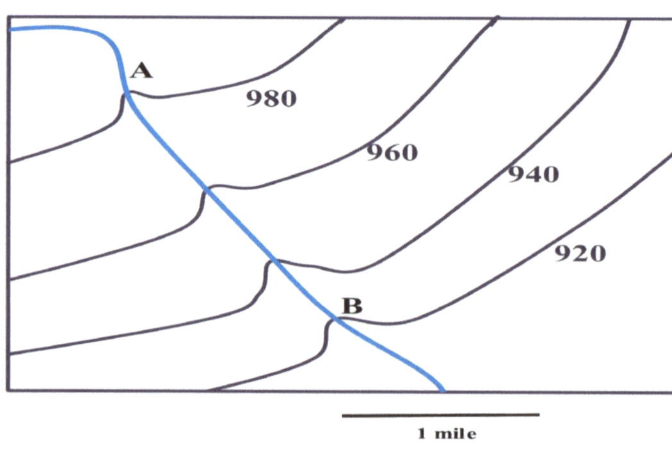

Figure 5.4 | Gradient calculation.

The elevation of the stream at A is 980', and the elevation of the stream at B is 920'. Use the scale bar to calculate the distance from A to B. Gradient = (980' – 920') / 2 miles, or 30 feet/mile. Stream gradients tend to be higher in a stream's headwaters (where it originates), and lower at their mouth, where they discharge into another body of water (such as the ocean). Discharge quantifies stream flow at a specified time and location, specifically measuring the amount of water traversing a designated point within a defined time interval. The discharge is calculated by multiplying the size of the stream channel (width multiplied by depth) by the water velocity, typically expressed in cubic feet (or meters) per second. Discharge escalates downstream in the majority of rivers when tributaries converge with the primary channel, contributing additional water.

The sediment load, or the quantity of sediment transported by the stream, varies from the headwaters to the outlet. At the headwaters, tributaries swiftly transport their sediment downstream, merging with contributions from additional tributaries. The primary river subsequently deposits its sediment load upon reaching base level. Occasionally, during the transportation of materials downstream, the sediment load becomes sufficiently substantial that the water cannot sustain it, resulting in deposition. When a stream becomes saturated with sediment, braided streams may form, characterized by a network of intersecting channels that mimics braided hair. Sand and gravel bars are characteristic of braided streams, prevalent in dry and semiarid environments with elevated erosion rates. Straight streams, characterized by their virtually linear channels, are less frequently observed, resulting from a linear zone of weakness in the underlying geology. Artificial straight channels may be constructed for the purpose of flood management.

Streams may exhibit meandering patterns characterized by broadly looped bends that resemble "S"-shaped curves. The swiftest water in a meandering stream flow from outer bend to outside curve. The increased velocity and turbulence result in enhanced erosion on the outer bend, creating a formation known as a cut bank. Erosion on one bank is counterbalanced by deposition on the opposing bank of the stream, where the slower-moving water facilitates sedimentation. These accumulations are referred to as point bars.

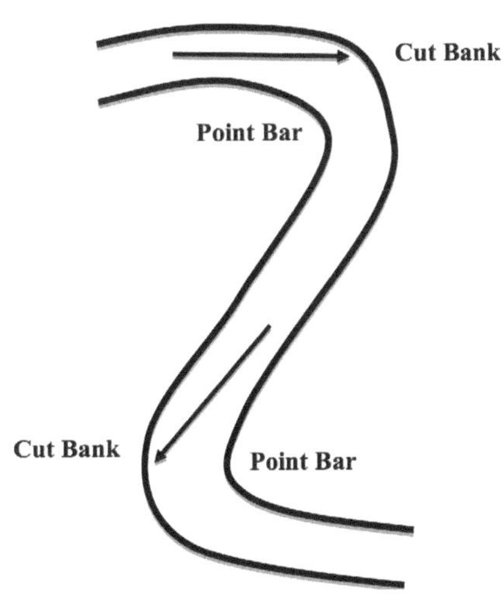

Figure 5.5 | Parts of a meandering stream. The S-curves are meanders. The arrows within the stream depict where the fastest water flows. That water erodes the

outside bank, creating a steep bank called the cut bank. The slowest water flows on the inside of the meander, slow enough to deposit sediment and create the point bar.

As meanders get increasingly intricate or sinuous, they may sever a meander, resulting in the formation of a crescent-shaped oxbow lake. Refer to Figure 5.6 to observe the creation of an oxbow lake.

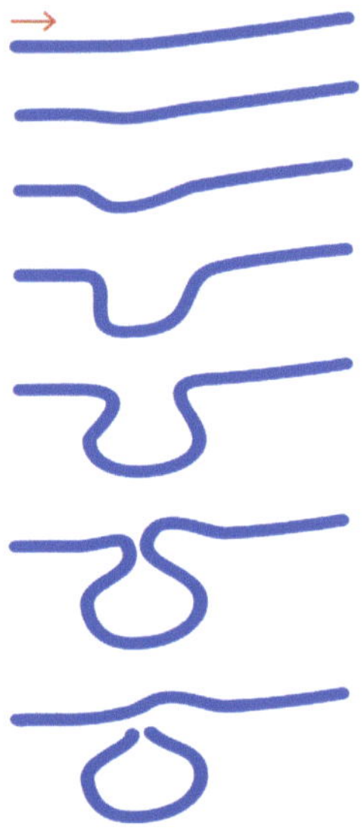

Figure 5.6 | Formation of an oxbow lake. A meander begins to form and is cut off, forming the oxbow.

Although streams are inanimate, they undergo distinctive transformations over time as they alter the landscape. The primary objective of a stream is to attain base level, defined as the lowest elevation at which the stream ceases to erode its course, typically a lake or another stream; the ultimate base level is the ocean. In pursuit of this objective, the stream will undergo the process of stream erosion, comprising the following steps.:

- Youthful (early) stage – these streams are downcutting their channels (vertically eroding); literally they are picking up sediment from the bottom of their channels in an effort to decrease their elevation. The land surface will be above sea level, and these streams form deep V-shaped channels.
- Mature (middle) stage – these streams experience both vertical (downcutting) and lateral (meandering) erosion. The land surface is sloped, and streams begin to form floodplains (the flat land around streams that are subject to flooding).
- Old age (late) stage – these streams focus on lateral erosion and have very complicated meanders and oxbow lakes. The land surface is near base level.

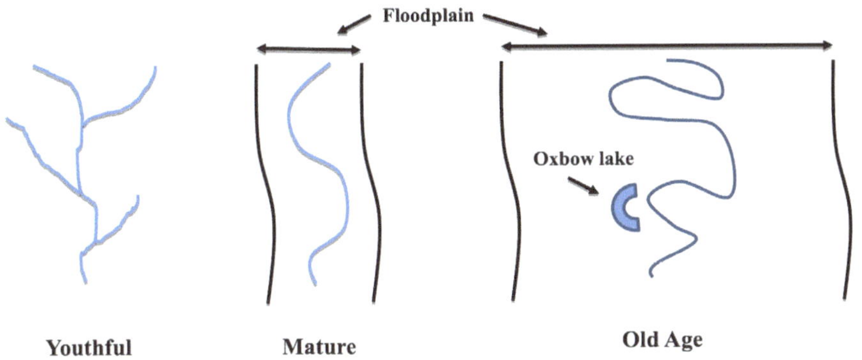

Figure 5.7 | Streams displaying the youthful, mature, and old age stages within the cycle of stream erosion. Note that the youthful stream does not have a floodplain.

An interruption may occur in this cycle. If a stream suddenly begins to downcut again, if sea level dropped (so base level dropped) or if the area around it was uplifted (think building mountains), then the stream would become rejuvenated. If the rejuvenated stream was in the old age stage, it will begin to form a deep V-shaped channel within that complicated meandering pattern that it has. This creates a neat geologic feature called an entrenched meander (Figure 5.8).

Figure 5.8 | Entrenched meanders along the San Juan River, Goosenecks State Park, Utah.

5.3.1 LAB EXERCISE

Part A – Drainage Patterns

Map 5.1, located on the next page, is a portion of a topographic map from Royal Gorge, Colorado, courtesy of the USGS. Study the map and answer the questions below.

1. What type of stream drainage pattern is present on this map? This may be easier to determine by examining the tributaries to the main stream.

   a. dendritic b. trellis c. radial d. rectangular

2. Based on the drainage pattern type, does the bedrock underlying this area consist of rocks uniformly resistant to erosion or rocks alternating between resistant and non-resistant layers?

   a. uniformly resistant bedrock b. alternately resistant and non-resistant bedrock

3. Are the rocks likely tilted and folded or horizontal?

   a. tilted and folded b. horizontal

4. Are streams in this area downcutting or laterally eroding?

   a. downcutting b. laterally eroding

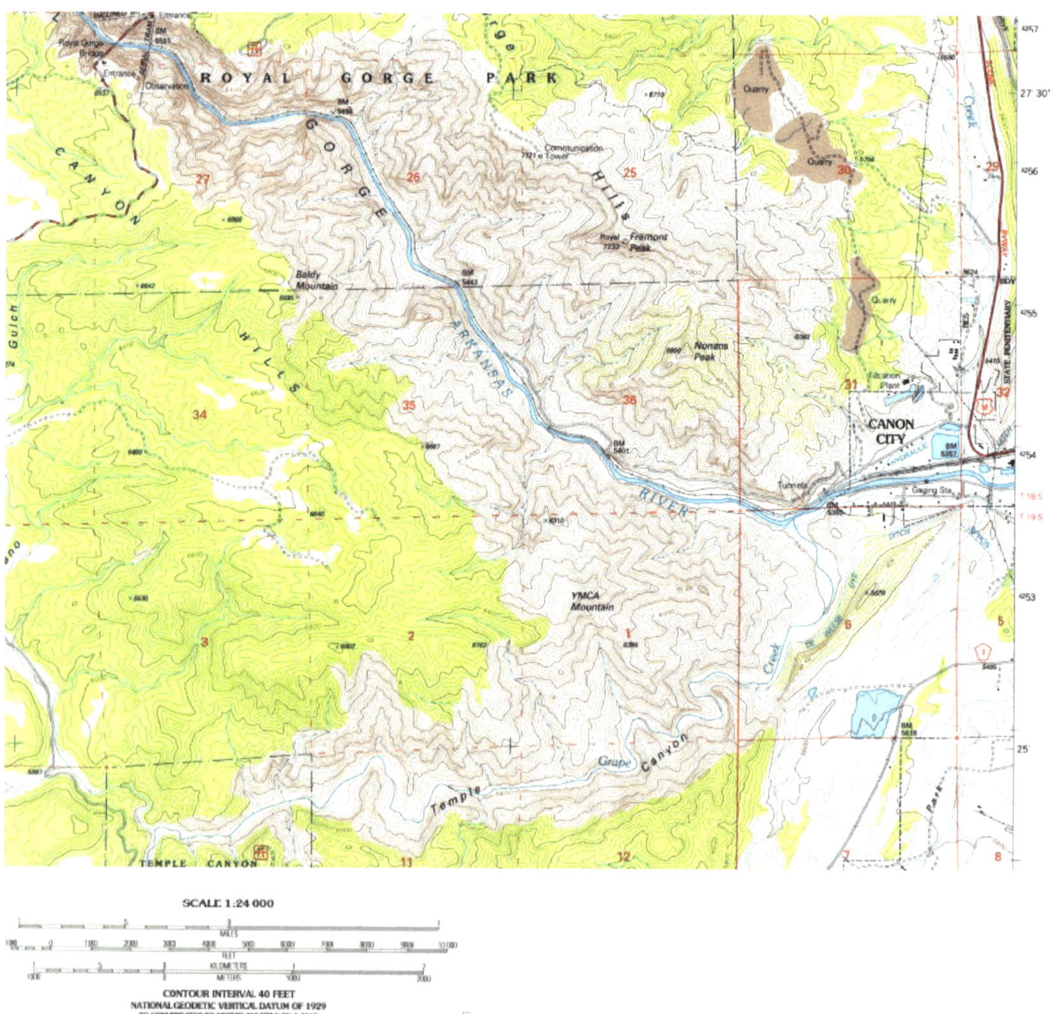

Map 5.1 | Royal Gorge, Colorado 1:24,000 topographic map

5. In which stage of the cycle of stream erosion is this area?

   a. old age b. mature c. youthful

Map 5.2 is the southeast portion of the map above, magnified to show Grape Creek. Use this map to answer question 6.

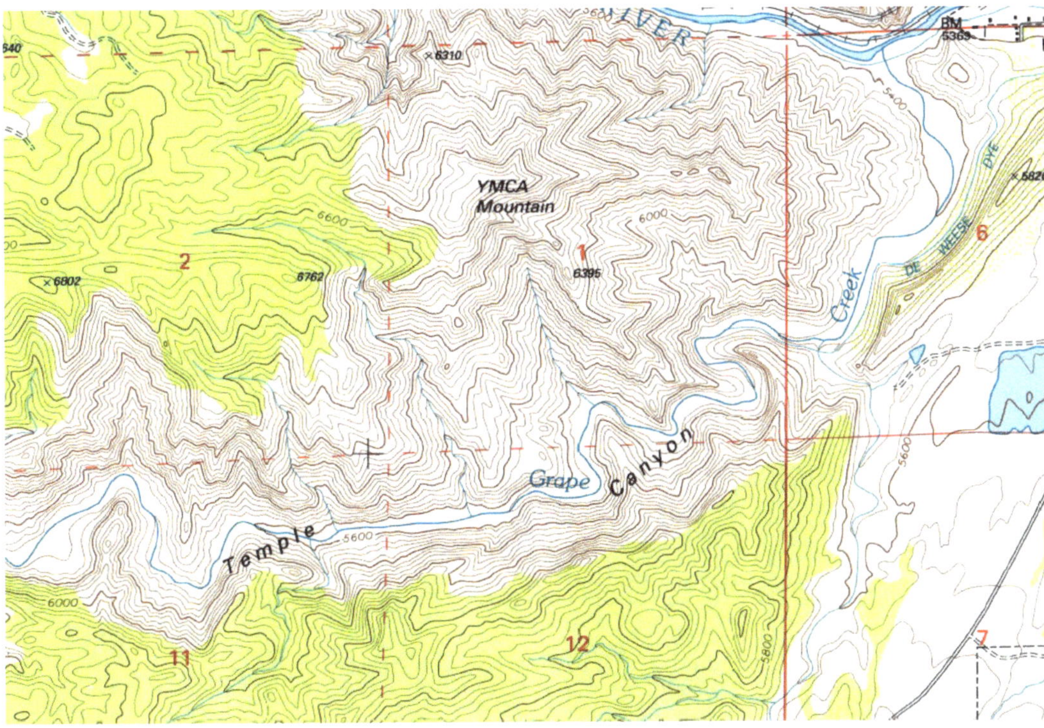

Map 5.2 | A portion of the Royal Gorge, Colorado map, zoomed in at Grape Creek

6. Calculate the stream gradient of Grape Creek in ft/mile. Use the index contour just above the "m" in Temple Canyon for your initial spot. Measure the gradient to the index contour past the word Creek (the last contour before it reaches Arkansas River). The distance between these areas is ~1.6 miles (measured along the curving distance of the non-magnified stream). What is the gradient?

   a. 15'/mile b. 100'/mile c. 125'/mile d. 200'/mile

Use the Omaha North, Nebraska-Iowa Map 5.3 on the next page to answer questions 7-11.

7. Observe the stream on the Omaha N, Nebraska-Iowa quadrangle. In which stage of the cycle of stream erosion is this area?

   a. old age b. mature c. youthful

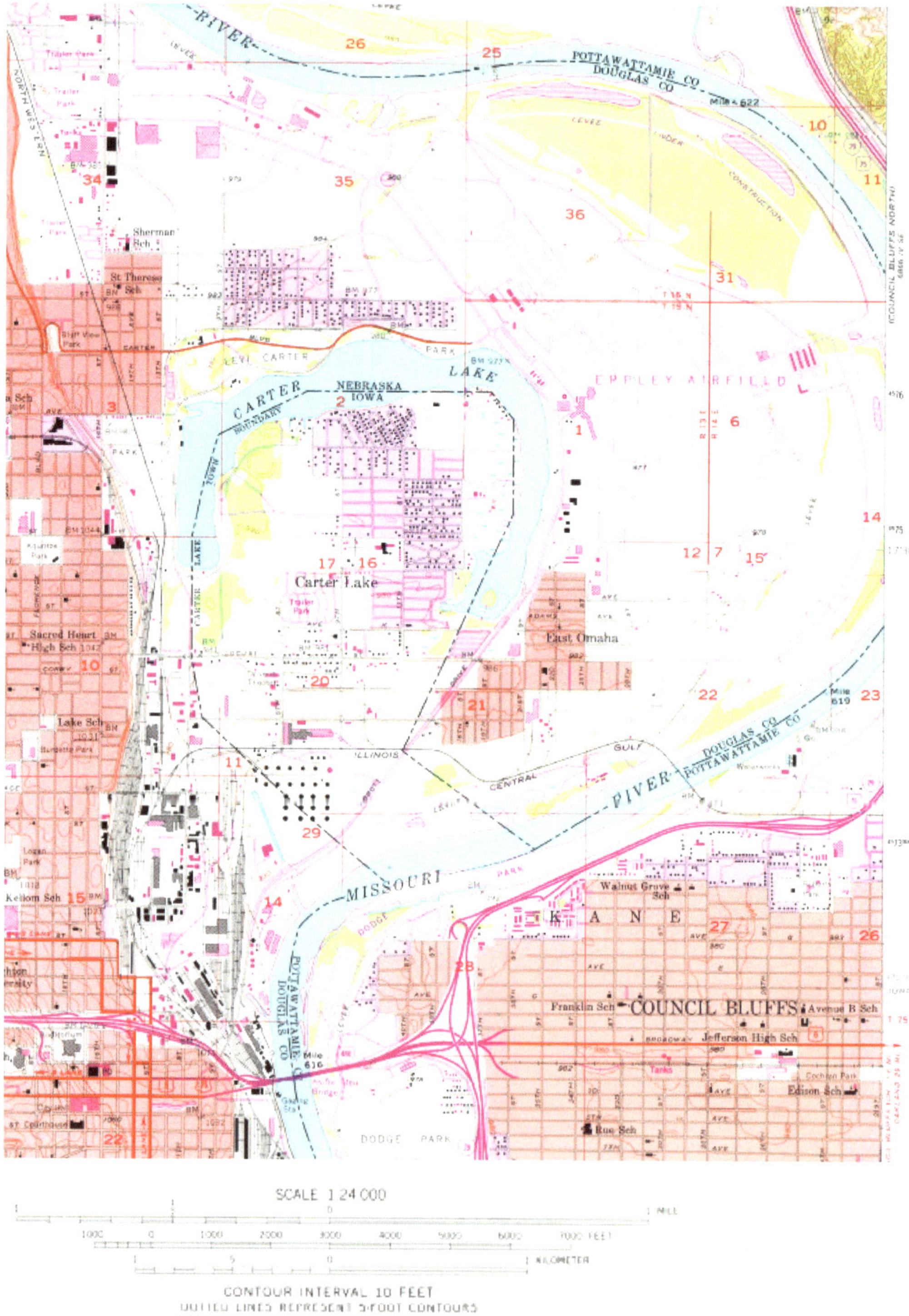

Map 5.3 | Omaha North,Nebraska-Iowa 1:24,000 topographic map

8. Compare the contour intervals from the Royal Gorge, Colorado map (map 5.1) to the Omaha N, Nebraska-Iowa map. Would you expect the gradient of the Missouri River in Nebraska to be greater or less than the gradient that you calculated for the Grape River in Colorado?

   a. less than the Grape River gradient b. greater than the Grape River gradient

9. Locate the state boundary between Nebraska and Iowa along the Missouri River. Why does the boundary depart from the river channel?

   a.when the boundary was created, Iowans wanted the Carter Lake area in their state

   b.the boundary follows the course of the river at the time that it was drawn; the river has since moved

   c.none of the above

10. What is the term for the geologic feature called Carter Lake?

    a. entrenched meander b. oxbow lake

    c. cutbank d. point bar

11. Was Carter Lake cut off before or after the state boundary between Nebraska and Iowa was drawn?

    a. before b. after

5.4 FLOODING

Flooding is a prevalent and significant issue in the rivers of our nation. Flood stage occurs when the water level in a stream surpasses its banks. Floodplains are desirable locations for construction due to their appealing water views; nonetheless, they are more suitably designated for recreational areas such as playgrounds and golf courses. Have you ever encountered the phrase "a 1 in 100 year flood"? Does this imply that a flood of comparable proportions will transpire every century? No, it merely indicates that, on average, we can anticipate a flood of this magnitude or bigger to transpire within a 100-year timeframe. It is impossible to ascertain that a flood of such scale will transpire in a certain year; rather, each year possesses a 1 in 100 probability of seeing such an event. This does not imply that only one flood of that magnitude can transpire within a century.

Figure 5.9 | USGS officials monitoring the flooding of Sweetwater Creek over I-20 in September, 2009.

To enhance comprehension of stream dynamics, the U.S. Geological Survey has deployed thousands of stream gauges nationwide, which serve as permanent water level indicators and recorders. Data from these stations can be utilized to construct flood frequency curves, which are crucial in informing flood management decisions. In September 2009, a catastrophic flood event transpired in western Georgia, leading to 11 fatalities, over $200 million in property damage, and the closure of Interstate 20 for one day. Precipitation occurred from September 16 to 22, with a notably intense episode on September 20. Utilize the data provided from a stream gauge situated on Sweetwater Creek near Austell, Georgia, to construct a flood frequency graph.

5.4.1 LAB EXERCISE

Part B – Recurrence Intervals

Data from the chart below was collected at the USGS site, and includes the 20 largest discharge events for Sweetwater Creek at station 02337000 from January 1, 2008 – May 1, 2015, excluding the dramatic 2009 flood (we will learn more about it later). In order to create a flood frequency graph, first the recurrence interval must be calculated (one is calculated below for an example). A **recurrence interval** refers to the average time period within which a given flood event will be equaled or exceeded once. To calculate it, first determine the rank of the flood, with a 1 going to the highest discharge event and a 20 going to the lowest discharge event. Calculate the recurrence interval using the following equation:

$$RI = (n+1) / m$$

- where RI = Recurrence Interval (yrs)
- n = number of years of record (in this case, 8)
- m = rank of flood

Peak Discharge Date	Discharge (cfs – cubic ft/sec)	Rank	Recurrence Interval
8/27/2008	5,140		
3/2/2009	2,360		
10/13/2009	3,290		
11/12/2009	6,120	1	9
12/3/2009	2,860		
12/10/2009	2,170		
12/19/2009	3,830		
Peak Discharge Date	Discharge (cfs – cubic ft/sec)	Rank	Recurrence Interval
12/26/2009	2,650		

1/25/2010	2,500		
2/6/2010	3,680		
3/12/2010	3,600		
3/10/2011	2,350		
4/17/2011	3,100		
2/24/2013	2,060		
2/27/2013	2,190		
5/6/2013	3,610		
12/23/2013	3,790		
4/8/2014	4,170		
1/5/2015	3,970		
4/20/2015	2,940		

Now that you have completed the chart, plot the discharge against the recurrence interval on the following graph. Please note that the x-axis (for recurrence interval) is in logarithmic scale and you may need to estimate where the data points fall. A logarithmic scale is <u>*non-linear*</u>, based on orders of magnitude. Each mark on the x-axis is the previous mark multiplied by a value. Use a straight edge to draw a best fit line (a straight line along the graph that shows the general direction that the group of points seem to be heading – it doesn't have to hit every point on the graph) through the graph when you are done, and then answer the following questions. Make sure your best fit line continues to the end of the graph.

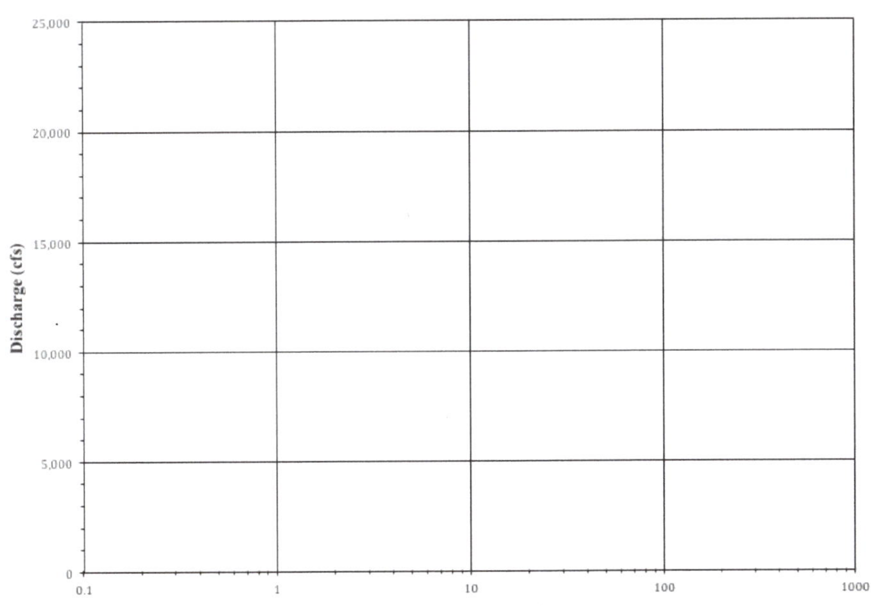

12. On which date did a flood event have a recurrence interval of 0.5?

    a. 2/27/2013 b. 10/13/2009 c. 3/10/2011 d. 4/20/2015

13. Of the following dated flood events, which one would you expect to happen more often?

    a. 8/27/2008 b. 2/24/2013 c. 2/6/2010 d. 12/23/2013

14. Observe your best fit line. What *approximate* discharge would be associated with a 50 year recurrence interval?

    a. 2,000 cfs b. 4,750 cfs c. 8,500 cfs d. 14,000 cfs

15. Flood stage, or bankfull stage, on Sweetwater Creek occurs at a discharge of ~4,500 cfs. According to your best fit line, what is the recurrence interval of such a discharge?

    a. 0.5 years b. 3 years c. 25 years d. 50 years

16. During the flood event of 9/23/2009, the discharge measured at this gaging station was 21,200 cfs. Note where this would plot on your graph. Would the recurrence interval for this flood plot at:

    a. 100 years b. 300 years c. 700 yearsd. longer than 1,000 years

17. Is it possible that a flood with a similar discharge to that of the event from 9/23/2009 could happen again in the next 20 years?

    a. Yes b. No

5.5 GROUNDWATER

It is preferable to avoid conceptualizing groundwater as subterranean lakes and streams, which are rarely found in caves; rather, consider groundwater as gradually permeating from one minute pore in the rock to another. Have you ever excavated a hole at the beach, only to have it inundated with water from below? If so, you had encountered the water table, the demarcation between the unsaturated and saturated zones. The rocks and soil located close below the land's surface constitute the unsaturated zone, wherein the pore spaces are occupied by air. Upon reaching the water table, rocks and soil pore spaces become inundated with water in the saturated zone.

The water table is described as reflecting topography, typically situated close to the ground surface, frequently tens of feet beneath it, but this can vary significantly by region. The water table ascends in elevated areas and descends in depressions, frequently draining into streams. The water table is augmented by inputs from rainwater that permeates the soil, referred to as recharge. The water table's location is dynamic; it decreases during droughts and increases during periods of precipitation.

Two critical qualities of groundwater that affect its availability and flow are porosity and permeability. Porosity denotes the open or vacant space contained within the rock. It is articulated as a proportion of the volume of voids relative to the overall rock volume. Porosity will differ according to rock type. Numerous rocks characterized by tightly interlocking crystals, such as igneous and metamorphic varieties, exhibit limited porosity due to the absence of free spaces. Sedimentary rocks consisting of well-sorted sediment typically exhibit high porosity due to the ample gaps between the constituent grains. Visualize a room entirely occupied by basketballs, akin to a rock formed entirely of sand grains. Introduce water into the room. The space can accommodate a substantial volume of water, as the basketballs do not fit closely together due to their configuration. This exemplifies great porosity..

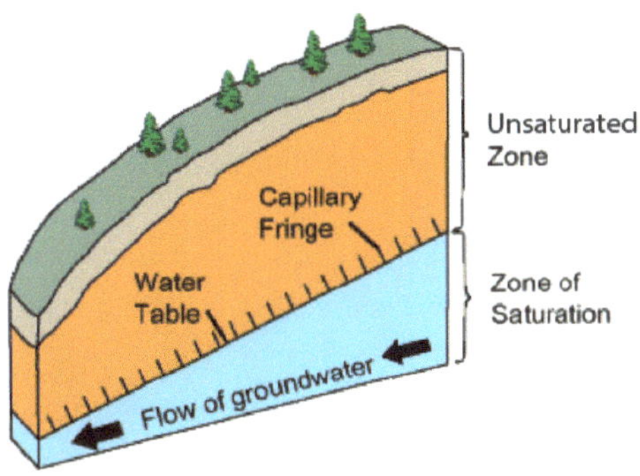

Figure 5.10 | The water table is the boundary between the unsaturated zone and saturated zone. Author: USGS

Permeability denotes the capacity of a geological substance to convey fluids. The factors include the rock's porosity, the dimensions of the voids, and the degree of interconnectivity among those voids. Despite a material's porosity, if the interstices are not interconnected, water will not permeate it.

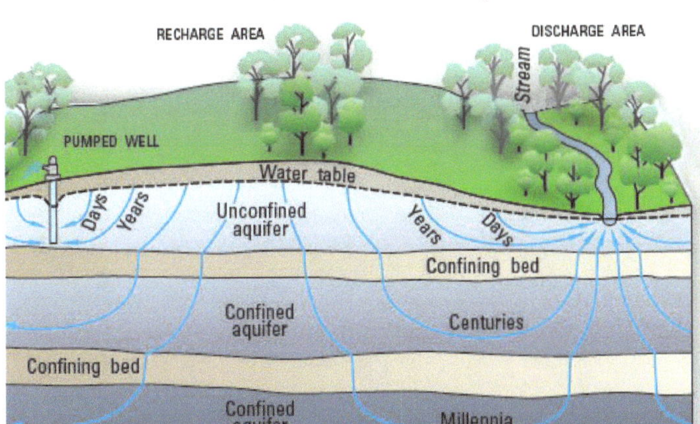

Figure 5.11 | Groundwater flow in both confined and unconfined aquifers.

Permeable rocks constitute effective aquifers, geological formations capable of yielding substantial water. Sedimentary rocks, including sandstone and limestone, serve as effective aquifers. Impermeable rocks create limiting layers that obstruct water movement. Confining layers include sedimentary rocks such as shale, composed of minute clay and silt particles, as well as unfractured igneous or metamorphic rock. The upper boundary of an unconfined aquifer is known as the water table.

Groundwater typically migrates from regions of elevated topography to those of lower topography within the shallow subsurface. Observe the flow routes depicted in Figure 5.11. About 20% of the water utilized in the United States is groundwater, which is susceptible to contamination mostly from sewage, landfills, industrial activities, and agriculture. The migration of groundwater facilitates the dispersion of pollutants, rendering containment a formidable challenge.

5.5.1 LAB EXERCISE

Part C – Groundwater Flow

Many gas stations use underground storage tanks (UST) to store fuel below the ground (you have likely seen a tanker truck at a gas station filling up the UST). These UST's could leak, and gasoline could possibly reach the water table. In the diagram below, a business using a well has detected gasoline in their groundwater. To detect the source of the potential leak, contour the water table's surface and determine its flow path. There are several gas stations in the diagram, and each has the potential to have the leaking UST. Seven monitoring wells are installed in the area, and you have data about the water table elevation within each well. Using that data (in elevation above sea level), contour this map as you would any other. Add the water table elevations to the map, and using pencil, contour the groundwater elevations using a contour interval of 2 feet. Noting that the gas should flow with the groundwater, determine the direction of groundwater flow and note the most likely gas station to be the source of the gasoline leak (in a real world scenario, once the likely culprit was determined, more monitoring wells would be installed and they would be tested for gasoline residue).

Table 5.1

Monitoring Well	Water table elevation (feet)
1	794'
2	790'
3	788'
4	786'
5	786'
6	783'
7	780'

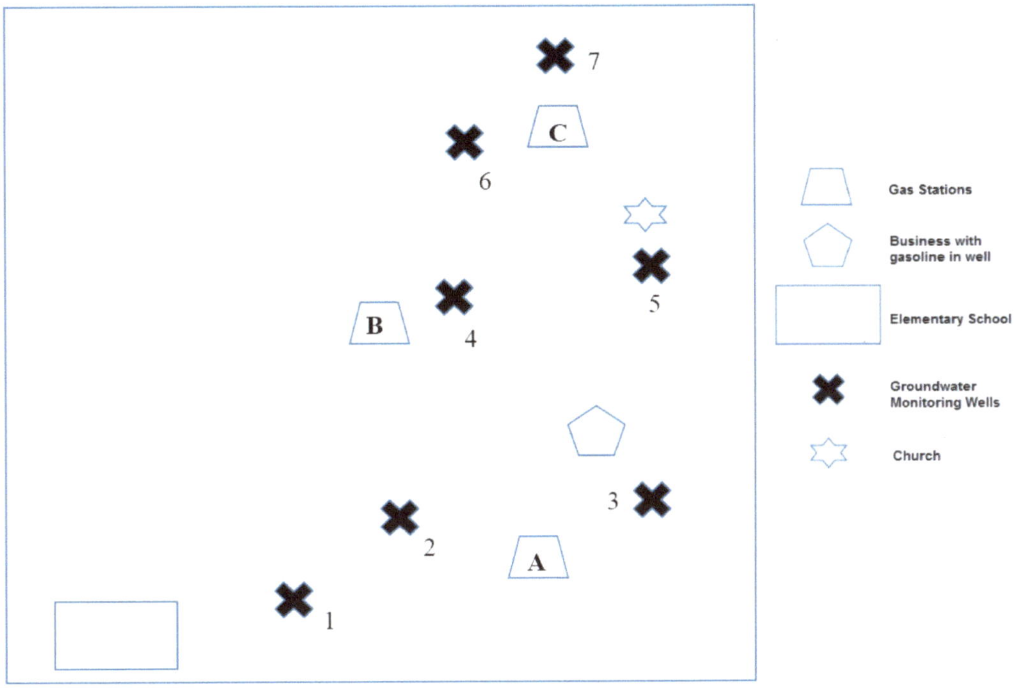

Figure 5.12 | Groundwater contamination exercise.

18. Which gas station is the most likely source of the gasoline leak?

    a. Station A   b. Station B   c. Station C

19. Is the school likely to be at risk of contamination from this same leak?

    a. Yes   b. No

20. Is the church likely to be at risk of contamination from this same leak?

    a. Yes   b. No

5.6 KARST TOPOGRAPHY

Limestone, a sedimentary rock, consists of the mineral calcite, which is soluble in water, particularly when the water is mildly acidic. In humid regions containing limestone, water erodes the rock, creating extensive caverns and depressions that differ in size and shape. Increased dissolution leads to cave instability and subsequent collapse, resulting in sinkholes. These expansive, crater-like depressions exemplify karst geology, called after the Karst area in Slovenia. Karst topography is defined by the presence of sinkholes, sink lakes (sinkholes containing water), caves, and disappearance streams (surface streams that vanish into a sinkhole). Inhabitants of karst topography face many obstacles, with around one-quarter of Americans in the contiguous 48 states residing in these areas. Sinkholes can emerge swiftly and inflict significant harm on any structures situated above them.

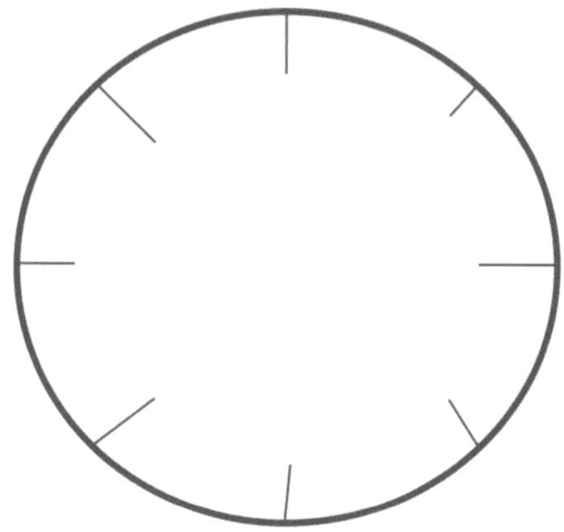

Figure 5.13 | Appearance of a sinkhole on a topographic map.

5.6.1 LAB EXERCISE

Part D – Karst topography

Use the Mammoth Cave, Kentucky map (Map 5.4 located at the end of this chapter) to answer the following questions. Though it is old (1922), the general geology and landforms in this area have not changed.

21. Locate Little Sinking Creek in the southern portion of the map, north of Hwy. 68 and south of the Edmonson County Line. In which direction does it flow?

    a. south b. north c. southeast d. northwest

22. Follow the creek along its path. Where does it wind up?

    a. Along Hwy. 68 b. It disappears underground

23. Find Sloans Crossing. It is south of Mammoth Cave. What is the benchmark elevation at Sloans Crossing?

    a. 600' b. 630' c. 800' d. 834'

24. Now look farther south of Sloans Crossing at Hwy. 31W. Look closely at the topography south of highway, and it changes abruptly. What feature(s) can you observe south of Hwy. 31?

    a. sinkholes

    b. disappearing streams

    c. generally lower land surface elevations

    d. all of the above

25. Keeping the abrupt topography change in mind, which of the following is true?

    a. In the northern portion of the map, the area is underlain by limestone

    b. In the southern portion of the map, the area is underlain by limestone

26. Locate the Louisville and Nashville Railroad line just south of Hwy. 31. Would this be an easy location to maintain a railroad?

    a. Yes b. No

Part E – Google Earth

The exercises that follow use Google Earth. For each question (or set of questions) paste the location that is given into the "Search" box. Examine each location at multiple eye altitudes and differing amounts of tilt. For any measurements that use the ruler tool, this can be accessed by clicking on the ruler icon above the image.

Search for 7 26 16.49S 75 00 00.06W and zoom out to an eye altitude of 15 miles.

27. How would one describe this river?

    a. Straight b. Meandering c. Low sinuosity d. Braided

28. In this stream, erosion is occurring on the _________ because_______ while deposition is occurring on the______because ______.

    a.point bars; the fastest velocity water flows to this point; cut banks; the slowest velocity water flows to this point

    b.point bars; the slowest velocity water flows to this point; cut banks; the fastest velocity water flows to this point

    c.cut banks; the fastest velocity water flows to this point; point bars; the slowest velocity water flows to this point

    d.cut banks; the slowest velocity water flows to this point; point bars; the fastest velocity water flows to this point

Search for 63 55 55.23N 17 01 07.14W and zoom out to an eye altitude of 10,000 feet.

29. How would one describe this river?

    a. Straight b. Meandering c. Low sinuosity d. Braided

30. What factors control the course of this river?

    a.Steep gradient and high discharge

    b.Low gradient and low discharge

    c.Low gradient and abundant sediment supply

    d.Steep gradient and low sediment supply

Search for 38 01 12.18N 121 43 20.02W and zoom out to an eye altitude of 30,000 feet.

31. The river in this area has a rather particular pattern, what geologic process caused this?

    a. a meander eroded through its bank and created an oxbow lake

    b. the river is in a karst terrain and disappeared into the ground

    c. the river is following patterns, likely faults, in the underlying bedrock

    d. during a flood the river breached the natural levee flowing into the floodplain

32. Zoom out and examine the surrounding area, what geological hazards are likely in the area?
    a. sinkholes
    b. flooding of urban areas
    c. erosion and subsidence
    d. none of the above

Search for 41 24 30.77N 122 11 46.23W and zoom to an eye altitude 50,000 feet.

33. What type of drainage pattern is present in this area?
    a. Trellis b. Dendritic c. Rectangular d. Radial e. Deranged
34. What does this type of drainage pattern indicate about the area?
    a. rocks in the area are homogeneous and/or flat lying
    b. rocks in the area are alternating resistant and non-resistant, forming parallel ridges and valleys
    c. stream channels radiate outward like wheel spokes from a high point
    d. stream channels flow randomly with no relation to underlying rocks or structure

Search for 36 45 41.23N 85 18 34.22W and zoom out to an eye altitude of 25,000 ft. Although there is no stream present today, the stream channels can be clearly seen.

35. What type of drainage pattern was present in this area?
    a. Trellis b. Dendritic c. Rectangular d. Radial e. Deranged
36. In what direction was the main river flowing?
    a. west b. east c. north d. south

Search for 28 38 01.92N 81 22 44.78W and zoom out to an eye altitude of 13,000 ft.

37. How were these lakes formed?
    a. they were man-made – all are dammed
    b. they are formed by large rivers in the area
    c. as sinkholes, as underlying soluble rock was dissolved and areas collapsed
    d. they are impact structures that filled with water
38. What type of bedrock is present in this area?
    a. limestone b. sandstone c. gneiss d. granite e. chert

Now let's travel to Mars! Mars has shown evidence of water, and NASA and others have been studying Mars intensely in recent years. Let's look at a few features on Mars. In the location toolbar across the top of Google Earth, locate the image that looks like the planet Saturn. Click the button to bring up several location options and select Mars. In the Layers box on the left, notice Global Maps. You may have to click on it to expand it. For today, we will use the highest quality images found in the Visible Images layer. The other layers are interesting, but won't be used today.

In the Search tab, type in Noctis Labyrinthus and zoom to ~300 miles.

39. These features you are seeing are linear valleys. Assume that water flowed through these valleys at some time. What type of drainage pattern would this area represent?

    a. Trellis b. Dendritic c. Rectangular d. Radial e. Deranged

40. Think about the drainage pattern you selected in the previous answer – what does this tell you about the underlying rocks?

    a. the rocks are probably fractured

    b. the rocks are uniformly resistant

    c. the rocks are part of a topographic high, like a mountain

    d. the rocks are alternately resistant and non-resistant

In the Search tab, type in Warrego Valles and zoom to ~120 miles.

41. Notice the general shape of this feature. It is thought to have formed by the runoff of either precipitation or groundwater. What type of drainage does this appear to be?

    a. Trellis b. Dendritic c. Rectangular d. Radial e. Deranged

5.7 STUDENT RESPONSES

1. What type of stream drainage pattern is present on this map? This may be easier to determine by examining the tributaries to the main stream.

   a. dendritic b. trellis c. radial d. rectangular

2. Based on the drainage pattern type, does the bedrock underlying this area consist of rocks uniformly resistant to erosion or rocks alternating between resistant and non-resistant layers?

   a. uniformly resistant bedrock b. alternately resistant and non-resistant bedrock

3. Are the rocks likely tilted and folded or horizontal?

   a. tilted and folded b. horizontal

4. Are streams in this area downcutting or laterally eroding?

   a. downcutting b. laterally eroding

5. In which stage of the cycle of stream erosion is this area?

   a. old age b. mature c. youthful

6. Calculate the stream gradient of Grape Creek in ft/mile. Use the index contour just above the "m" in Temple Canyon for your initial spot. Measure the gradient to the index contour past the word Creek (the last contour before it reaches Arkansas River).

The distance between these areas is ~1.6 miles (measured along the curving distance of the non-magnified stream). What is the gradient?

a. 15'/mile b. 100'/mile c. 125'/mile d. 200'/mile

7. Observe the stream on the Omaha N, Nebraska-Iowa quadrangle. In which stage of the cycle of stream erosion is this area?

   a. old age b. mature c. youthful

8. Compare the contour intervals from the Royal Gorge, Colorado map (map 5.1) to the Omaha N, Nebraska-Iowa map. Would you expect the gradient of the Missouri River in Nebraska to be greater or less than the gradient that you calculated for the Grape River in Colorado?

   a. less than the Grape River gradient

   b. greater than the Grape River gradient

9. Locate the state boundary between Nebraska and Iowa along the Missouri River. Why does the boundary depart from the river channel?

   a. when the boundary was created, Iowans wanted the Carter Lake area in their state

   b. the boundary follows the course of the river at the time that it was drawn; the river has since moved

   c. none of the above

10. What is the term for the geologic feature called Carter Lake?

    a. entrenched meander b. oxbow lake

    c. cutbank d. point bar

11. Was Carter Lake cut off before or after the state boundary between Nebraska and Iowa was drawn?

    a. before b. after

12. On which date did a flood event have a recurrence interval of 0.5?

    a. 2/27/2013 b. 10/13/2009 c. 3/10/2011 d. 4/20/2015

13. Of the following dated flood events, which one would you expect to happen more often?

    a. 8/27/2008 b. 2/24/2013 c. 2/6/2010 d. 12/23/2013

14. Observe your best fit line. What *approximate* discharge would be associated with a 50 year recurrence interval?

    a. 2,000 cfs b. 4,750 cfs c. 8,500 cfs d. 14,000 cfs

15. Flood stage, or bankfull stage, on Sweetwater Creek occurs at a discharge of ~4,500 cfs. According to your best fit line, what is the recurrence interval of such a discharge?

    a. 0.5 years b. 3 years c. 25 years d. 50 years

16. During the flood event of 9/23/2009, the discharge measured at this gaging station was 21,200 cfs. Note where this would plot on your graph. Would the recurrence interval for this flood plot at:

    a. 100 years b. 300 years c. 700 years d. longer than 1,000 years

17. Is it possible that a flood with a similar discharge to that of the event from 9/23/2009 could happen again in the next 20 years?

    a. Yes b. No

18. Which gas station is the most likely source of the gasoline leak?

    a. Station A b. Station B c. Station C

19. Is the school likely to be at risk of contamination from this same leak?

    a. Yes b. No

20. Is the church likely to be at risk of contamination from this same leak?

    a. Yes b. No

21. Locate Little Sinking Creek in the southern portion of the map, north of Hwy. 68 and south of the Edmonson County Line. In which direction does it flow?

    a. south b. north c. southeast d. northwest

22. Follow the creek along its path. Where does it wind up?

    a. Along Hwy. 68 b. It disappears underground

23. Find Sloans Crossing. It is south of Mammoth Cave. What is the benchmark elevation at Sloans Crossing?

    a 600' b. 630' c. 800' d. 834'

24. Now look farther south of Sloans Crossing at Hwy. 31W. Look closely at the topography south of highway, and it changes abruptly. What feature(s) can you observe south of Hwy. 31?

    a. sinkholes

    b.disappearing streams

    c. generally lower land surface elevations

    d. all of the above

25. Keeping the abrupt topography change in mind, which of the following is true?

    a. In the northern portion of the map, the area is underlain by limestone

    b. In the southern portion of the map, the area is underlain by limestone

26. Locate the Louisville and Nashville Railroad line just south of Hwy. 31. Would this be an easy location to maintain a railroad?

    a. Yes b. No

27. How would one describe this river?
 a. Straight b. Meandering c. Low sinuosity d. Braided

28. In this stream, erosion is occurring on the___ because_______,while deposition is occurring on the___because_______.
 a. point bars; the fastest velocity water flows to this point; cut banks; the slowest velocity water flows to this point
 b. point bars; the slowest velocity water flows to this point; cut banks; the fastest velocity water flows to this point
 c. cut banks; the fastest velocity water flows to this point; point bars; the slowest velocity water flows to this point
 d. cut banks; the slowest velocity water flows to this point; point bars; the fastest velocity water flows to this point

29. How would one describe this river?
 a. Straight b. Meandering c. Low sinuosityd. Braided

30. What factors control the course of this river?
 a. Steep gradient and high discharge
 b. Low gradient and low discharge
 c. Low gradient and abundant sediment supply
 d. Steep gradient and low sediment supply

31. The river in this area has a rather particular pattern, what geologic process caused this?
 a. a meander eroded through its bank and created an oxbow lake
 b. the river is in a karst terrain and disappeared into the ground
 c. the river is following patterns, likely faults, in the underlying bedrock
 d. during a flood the river breached the natural levee flowing into the floodplain

32. Zoom out and examine the surrounding area, what geological hazards are likely in the area?
 a. sinkholes
 b. flooding of urban areas
 c. erosion and subsidence
 d. none of the above

33. What type of drainage pattern is present in this area?
 a. Trellis b. Dendritic c. Rectangular d. Radial e. Deranged

34. What does this type of drainage pattern indicate about the area?
 a. rocks in the area are homogeneous and/or flat lying

b. rocks in the area are alternating resistant and non-resistant, forming parallel ridges and valleys

c. stream channels radiate outward like wheel spokes from a high point

d. stream channels flow randomly with no relation to underlying rocks or structure

35. What type of drainage pattern was present in this area?

a. Trellis b. Dendritic c. Rectangular d. Radial e. Deranged

36. What does this type of drainage pattern indicate about the area?

a. rocks in the area are homogeneous and/or flat lying

b. rocks in the area are alternating resistant and non-resistant, forming parallel ridges and valleys

c. stream channels radiate outward like wheel spokes from a high point

d. stream channels flow randomly with no relation to underlying rocks or structure

37. In what direction was the main river flowing?

a. west b. east c. north d. south

38. How were these lakes formed?

a. they were man-made – all are dammed

b. they are formed by large rivers in the area

c. as sinkholes, as underlying soluble rock was dissolved and areas collapsed

d. they are impact structures that filled with water

39. What type of bedrock is present in this area?

a. limestone b. sandstone c. gneiss d granite e. chert

40. These features you are seeing are linear valleys. Assume that water flowed through these valleys at some time. What type of drainage pattern would this area represent?

a. Trellis b. Dendritic c. Rectangular d. Radial e. Deranged

41. Think about the drainage pattern you selected in the previous answer – what does this tell you about the underlying rocks?

a. the rocks are probably fractured

b. the rocks are uniformly resistant

c. the rocks are part of a topographic high, like a mountain

d. the rocks are alternately resistant and non-resistant

42. Notice the general shape of this feature. It is thought to have formed by the runoff of either precipitation or groundwater. What type of drainage does this appear to be?

a. Trellis b. Dendritic c. Rectangular d. Radial e. Deranged

Map 5.4 | (appearing on following page) Mammoth Cave, Kentucky 1:24,000 topographic map

6 Chapter

CLIMATE CHANGE

6.1 INTRODUCTION: AN OVERVIEW

Climate represents the average of long-term weather patterns in a certain geographic region, a complex measure influenced by elements within the lithosphere, atmosphere, cryosphere, hydrosphere, biosphere, and anthrosphere, as well as extraterrestrial variables. Distinguishing people from other forms of life (anthrosphere versus biosphere) is beneficial for various reasons, chiefly because many of our actions, such as industrialization, are unparalleled in the natural world, and this differentiation aids in comprehending our impact on climate change. Consequently, the study of historical, contemporary, and prospective climate is very intricate and multidisciplinary. Although climate may not be perceived as a geological discipline, its historical record is embedded in rocks, contemporary climate is modified by geological phenomena, and future climate will be affected by our use of geological resources, including fossil fuels. This issue is not only complex but also one of the most crucial scientific fields, essential for comprehending the dynamics and implications of future climate change and for efforts to resist or reduce its possible effects.

While the fundamental science of climate and climate change has been extensively researched and largely agreed upon among the scientific community, considerable controversy persists among the general populace. This is probably influenced by other issues beyond science, including economics, politics, media representation of science, and the general public's scientific knowledge. Acquiring a comprehensive understanding of this problem is challenging due to the vast amount of information and the variation in scientific literacy. This laboratory will investigate the issue by analyzing climate data and evaluating how we, as scientists or scientifically inclined individuals, evaluate and draw conclusions from data, its presentation, and its relevance to our comprehension of the surrounding world.

6.2 THE CLIMATE SYSTEM

As previously stated, climate refers to the long-term meteorological patterns of a location. It is crucial to underscore the long-term aspect of the definition to clarify that climate differs from weather. Weather refers to the localized and short-term variations in temperature, humidity, precipitation, air pressure, wind, and other meteorological factors. Weather varies over the day, week, month, and year, making it challenging to discern any trends beyond the inherent randomness of the system. A long-term perspective on weather reveals trends over time and geography that enhance our understanding of the variables influencing the climate system. The climate system is the interrelated network of factors that affect the Earth's climate, encompassing elements from the lithosphere, atmosphere, hydrosphere, cryosphere, biosphere, anthrosphere, and solar system.

The thermal energy that sustains this system originates from two principal sources. Initially, there is thermal energy emanating from the Earth, originating from the disintegration of radioactive substances and the remaining heat from the planet's birth. Heat distribution is uneven, with greater heat loss occurring in regions where the crust is thinner, such as near divergent borders. The Earth absorbs heat from solar energy. The distribution of heat throughout the Earth's surface is uneven, and the energy received correlates with the angle at which solar radiation strikes the Earth's surface (Figure 6.1). When solar energy strikes a surface perpendicularly, it is absorbed more effectively than at an oblique angle, which accounts for the higher temperatures in the tropics compared to the poles..

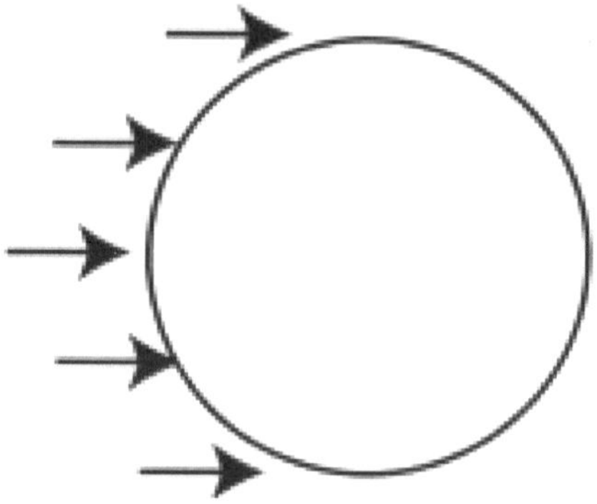

Figure 6.1 | The Earth's shape in-fluences the angle at which the sun's rays hit the surface from perpendicular at the equator to parallel at the poles. This creates large climate differences across the Earth.

The composition of the Earth's surface is significant as various materials respond differently to solar radiation. Certain materials, typically dark-hued, absorb and re-emit heat, with the majority being preserved near the planet's surface. If you have ever walked barefoot on dark concrete or asphalt during the summer, you are probably acquainted with this experience. Other materials that possess a glossy or bright hue reflect solar energy from the Earth's surface. Substances like snow or ice are especially proficient at reflecting sun light. Arctic explorers must utilize eye protection to prevent snow blindness. Albedo refers to the fraction of solar energy reflected from the Earth's surface, which can fluctuate based on the type of ground cover. The Earth's albedo is elevated when extensive areas are covered with glacial ice, resulting in reduced solar absorption and lower observed temperatures.

Once heat is reradiated from the Earth's surface, it ascends into the atmosphere. Specific gases in Earth's atmosphere, known as greenhouse gases, permit sunlight to enter but absorb terrestrial energy, subsequently radiating it in all directions, including back to the Earth's surface. Gases including water vapor, carbon dioxide, and methane constitute a minuscule yet significant portion of atmospheric composition. Various greenhouse gases differ in their efficacy to absorb and reradiate energy, as well as in their atmospheric concentrations, resulting in a greater retention of thermal energy with increased levels of potent greenhouse gases. The remaining reflected and radiated energy exits the atmosphere and disperses into space.

Various factors can affect the aforementioned simplified representation of the climate system. The solar radiation output fluctuates throughout time. The configuration of our orbit around the sun fluctuates throughout time from a more circular to a more elliptical form due to the gravitational effects of other planets within our solar system. The angle of incidence of solar radiation on the planet's surface is affected by the axial tilt and precession of the Earth. The distribution of water, ice, snow, vegetation, and other materials on the Earth's surface regulates its albedo and may vary over time. The concentration of greenhouse gases can fluctuate significantly based on the velocity of plate tectonics, the extent of volcanic activity, photosynthesis, rock weathering, fossil fuel combustion, and various other determinants. (to examine trends on climate and carbon dioxide levels see Figure 6.5 later in the lab).

The efficiency of the transportation of heat across the surface of the Earth also influences climate. Heat is transferred across the surface of the planet by wind, ocean currents, and storms. Therefore, the position of the continents as well as air and ocean currents affect climate and can change over time. The components in our atmosphere are also important, including water vapor and aerosols (dust). Water vapor in the atmosphere is a greenhouse gas, can reflect incoming solar radiation, and is the source of precipitation. Aerosols can come from the Earth's surface, ash from volcanoes, and the burning of fossil fuels and can alter climate by reflecting incoming solar radiation before it reaches the Earth's surface. There are numerous additional factors that also have some level of effect on the climate system.

At this point you should be able to recognize the complexity of the climate system based on the number of variables and how those variables can change over time. It is also important to recognize that all of these variables are connected. For instance, if a volcano erupts it adds some thermal energy to the climate system; it produces aerosols that block solar radiation from hitting the Earth's surface, and produces greenhouse gases that retain heat. Notice that these factors do not all influence climate in a consistent way. The change of one variable as the result of another is called feedback, which can be either positive or

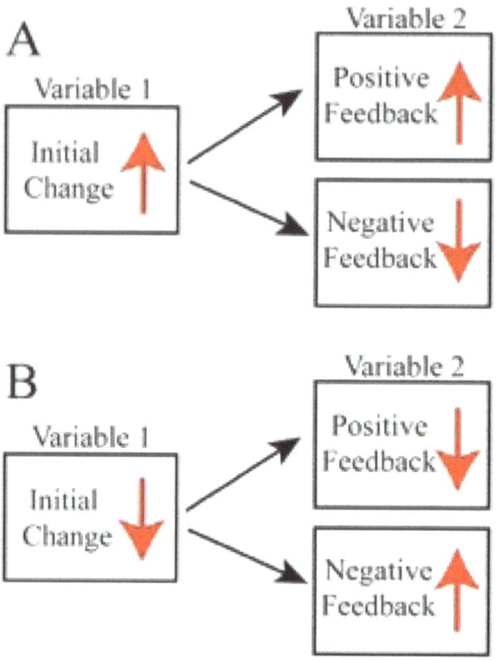

Figure 6.2 | A simple diagram showing the relationship between two variables with either positive or negative feedback following a positive (A) or negative (B) initial change.

(Figure 6.2). **Positive feedback** reinforces the initial change no matter the direction of that change. For instance, if the Earth warms, ice melts and reduces the albedo, which causes even more warming. This can also occur in the opposite direction, if the Earth cools, ice forms and increase the albedo, which causes more cooling. **Negative feedback** counteracts the initial change no matter the direction of the initial change. For instance, if the Earth warms, more area becomes arid resulting in an increase in the amount of dust in the atmosphere, which reflects solar radiation causing cooling. Conversely, if the Earth cools, a reduction in arid regions leads to a decrease in atmospheric dust, which subsequently induces warmth. Consequently, comprehending the climate system necessitates the identification of all significant climate variables, their interrelations, the rate of their potential changes, and the size and direction of alterations in each feedback loop. The optimal method to comprehend the climate system is to examine it via geological history.

6.3 CLIMATE PROXIES AND THE CLIMATIC RECORD

The first way that most students choose for tracking climate is employing a thermometer to directly gauge temperature. Reconstructing climatic patterns using this method has several challenges, notably that a thermometer provides a localized measurement and, more significantly, thermometers are a very modern innovation. Since direct observations fail to provide the long-term trends necessary to ascertain climate change or patterns, we must examine a natural recorder of climate known as a climate proxy. climatic change influences the deposition of sedimentary rock, rock chemistry, and fossil animals, which scientists analyze to reconstruct ancient climatic patterns in the discipline of paleoclimatology. A single climatic proxy may not provide a definitive indication of global climate for several reasons. Proxies reflect the historical context of their specific formation area, rather than that of an entire region. Secondly, an individual proxy, regardless of its extensive or limited history, can capture the short-term fluctuations of meteorological phenomena. Furthermore, the majority of climatic proxies are affected by several causes. The thickness of tree rings, as studied in dendrochronology, serves as an excellent proxy for temperature. Trees exhibit increased growth in warmer years, resulting in thicker rings, and diminished growth in colder years, leading to thinner rings. A tree may also have poor growth due to dry conditions or pest infestations, despite a warm year.

How can we rebuild long-term global temperature records if each individual proxy exhibits local patterns, accompanied by weather-related noise and potential influences from other factors? The solution resides in augmenting the dataset's size. If temperature is the primary variable affecting the proxies, and we aggregate hundreds to thousands of individual proxy data, a discernible pattern arises from the noise. An individual proxy record may contradict the overarching trend, which is anticipated, as a local area can experience a frigid winter during an otherwise warm year globally. To exemplify this, take the following: we aim to reconstruct overarching economic trends in the United States over the past few centuries. We might analyze other indicators of economic growth, including employment, the stock market, personal wealth, and home ownership rates, among others. Examining any one of these proxies would certainly obscure a comprehensive understanding of change. Furthermore, focusing solely on Macon, Georgia, would likely obscure a trend representative of the entire nation. A new factory established outside Macon, GA would significantly boost the local economy, but not the national economy as a whole. As the volume of data increases, whether pertaining to economics, climate, or any other system, the signal becomes increasingly discernible amidst local and random noise.

One of the most often utilized climate proxies is the assessment of oxygen isotopes. As noted in chapter one, isotopes are atoms of the same element that vary in mass due to changes in the number of neutrons within the atomic nucleus.

Figure 6.3 | Short segment of an ice core that records ancient climate patterns.

Multiple isotopes of oxygen are stable, meaning they do not radioactively decay over time. Oxygen has two stable isotopes that occur in a constant ratio on Earth. However, certain minerals (like calcite or ice) prefer one of the isotopes over the other within their crystal structure (a slightly larger or smaller atom fits better). This preference results in a ratio of oxygen isotopes that is different from the ratio found in other materials; this difference is called fractionation. The amount of fractionation in oxygen isotopes is temperature dependent, such that the mineral calcite has a different ratio of oxygen isotopes if it was formed in near freezing versus warm water temperatures. Using oxygen isotopes we can get climate records from many different sources, including coral, clams and other mollusks, the skeletons of single-celled organisms, and ice cores to name a few. Ice cores (as shown in Figure 6.3) can contain a wealth of climate data in addition to temperature data from oxygen isotopes, such as air bubbles that record the levels of greenhouse gases, concentrations of windblown aerosols, and ash from volcanic eruptions.

Other proxies include the extent of glacial sediment, sea level curves, pollen (palynology), and fossils. For instance, climatologists have used several features within fossil plants to reconstruct climate, largely because these organisms are sensitive to climate. These proxies include the thickness of tree rings, the shape of the leaves (toothier leaves are more common in colder climates), and the density of pores on leaf surfaces (more pores are needed with lower concentrations of carbon dioxide, which is necessary for photosynthesis).

As was mentioned before, by combining hundreds to thousands of individual climate records we can start to gain insight into overall climate trends. For instance, the Intergovernmental Panel on Climate Change (IPCC) and the National Oceanic and Atmospheric Administration (NOAA) regularly compile multiple types of proxy records from across the world (Figure 6.4) to reconstruct climate patterns (Figure 6.5).

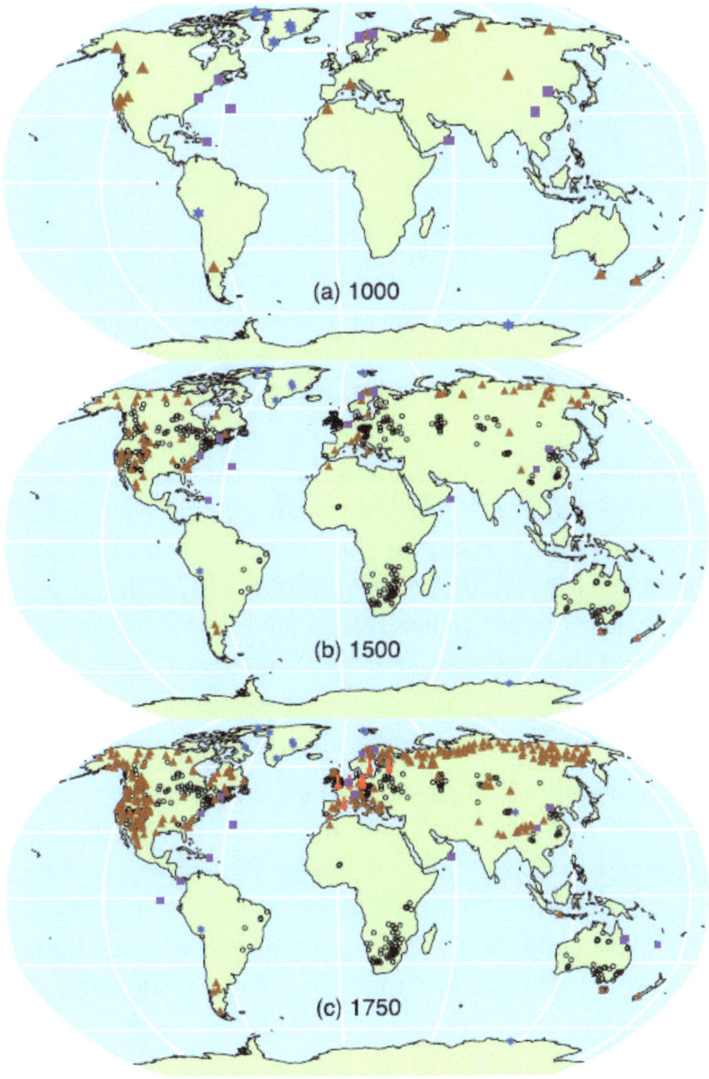

Figure 6.4 | Distribution of individual climate proxies used in the construction of Figure 5. Image from the National Ocean and Atmospheric Administration.

The accuracy of the climate records very much depends on the time frame being considered, with more certainty in the patterns of the recent past (Cenozoic) and less the further back in geologic time we are examining.

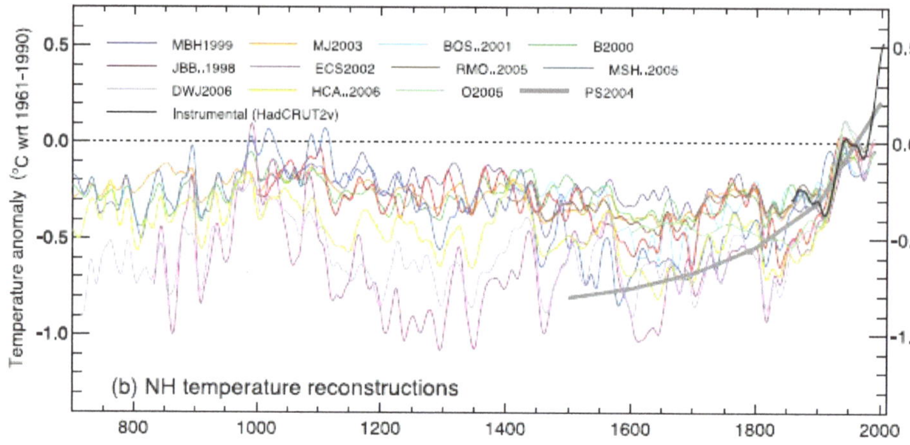

Figure 6.5 | Climate reconstruction over the last 1300 years using multiple climate proxies (different colored lines) from "Climate change 2007: the physical science basis"; Contribution of Working Group I to the Fourth Assessment Report of the Intergovernmental Panel on Climate Change Jansen, E. J. Overpeck, K. R. Briffa, J. Duplessy, F. Joss, V. Masson-Delmotte, D. Olago, B. Otto-Bliesner, W. R. Peltier, S. Rahmstorf, R. Ramesh, D. Raynaud, D. Rind, O. Solomina, R. Villalba, and D. Zhang.

The climate proxy we will focus on for this lab is the extent of sea ice coverage on the North and South Polar ice sheets. This is an easy proxy to assess from satellite images and is measured as the size of the ice sheet in million square kilometers. This proxy isn't a perfect indicator of global climate change, but it is easy to understand that a warming of the Earth is likely to cause a decrease in the amount of ice at the poles and thus a decrease in the **ice extent**, while a cooling event will cause an increase in ice production. Ice extent is simply the amount of geographic area covered by a glacier as measured from satellites.

The interpretation of individual proxies and the resultant temperature records is contentious, primarily due to the economic and political dimensions of climate change. This laboratory was established after to a dialogue with a student over the facts conveyed in several climate papers. The discourse centered on the presentation of scientific data to the public and the methodology for drawing conclusions from the supplied information. When evaluating information, it is essential to assess 1) the source of the data, 2) the methodology of data collection, 3) the manner of data presentation, and, most critically, 4) the rational conclusions that can be drawn from the data, independent of accompanying opinions.

6.3.1 LAB EXERCISE

As with Chapter 4, you will be expected to input your answers to this lab in several ways. There will be a couple of multiple-choice questions, but for the majority of the lab you will write your answers in the provided text box. This allows you to show your work in the questions requiring calculations as well as allowing you to answer open-ended questions thoroughly with multiple sentences. You will be expected to use correct grammar and complete sentences in your answers.

Materials

All of the data provided in the lab comes from the National Snow and Ice Data Center (NSIDC), housed at the University of Colorado. The data presented in this lab can be freely downloaded from them at nsidc.org. The original discussion was focused on a widely circulated though unattributed article entitled "Antarctic Sea Ice for March 2010 Significantly Greater Than 1980" published in April 2010 (climatechangehoax.com). For the dataset presented in this article as well as the following datasets, do the following before answering each set of questions: graph the data, connect the data points to better see the pattern through time, estimate the best fit line, and calculate the slope of the best fit line. Make sure to think about the data before you estimate the line of best fit so that your line falls within the data and is consistent with the trend in the data. An example of this process is shown in Figure 6.6.

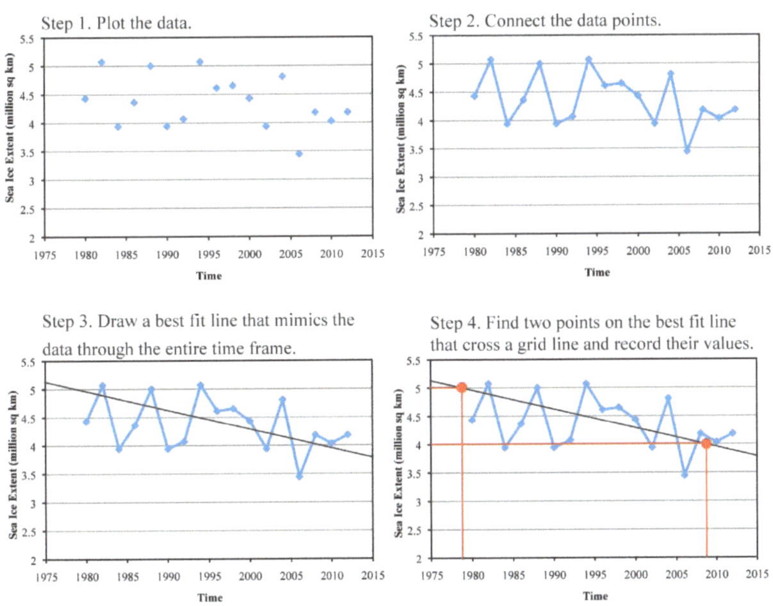

Step 5. Calculate the slope: $Slope = \frac{Rise}{Run}, which = \frac{y_2 - y_1}{x_2 - x_1}$

$$Slope = \frac{4\,million\,sq\,km - 5\,million\,sq\,km}{2009 - 1979} = \frac{-1\,million\,sq\,km}{30\,years} = -0.033\frac{million\,sq\,km}{year}$$

Figure 6.6 | Steps needed to analyze the ice extent data to visualize patterns: place a line of best fit, and calculate the slope of that line.

Part A – Original Data

First, we want to take a closer look at the data presented in the original article ("Antarctic Sea Ice for March 2010 Significantly Greater Than 1980") and interpret the data. Feel free to read the article, but it isn't needed to complete the assignment or understand the patterns it is presenting. The data, which is included to the left of the graph below, is the extent of Antarctic sea ice in millions of square kilometers as measured in March of 1980 and 2010. This data is accurate and is consistent with data that can be downloaded from NSIDC. To do this, follow the instructions in Figure 6.6. In this case, steps two and three are the same and you can easily calculate slope using the original data.

	Sea Ice Extent
2010	4.0 million square km
1980	3.5 million square km.

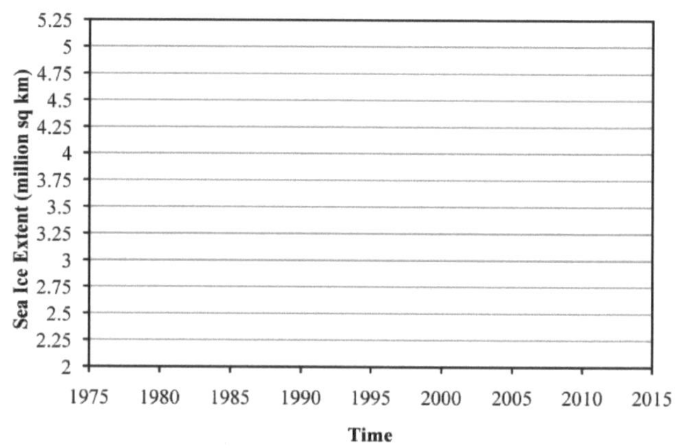

1. Based exclusively on the data provided and your graph, what conclusion would you make regarding climate change?
   a. Sea ice is expanding, which indicates an increase in temperature
   b. Sea ice is expanding, which indicates a decrease in temperature
   c. Sea ice is contracting, which indicates an increase in temperature
   d. Sea ice is contracting, which indicates a decrease in temperature
2. What is slope of the line of best fit for this data?
   a. 0.008 million square kilometers per year
   b. 0.05 million square kilometers per year
   c. -0.05 million square kilometers per year
   d. 0.017 million square kilometers per year
   e. -0.17 million square kilometer per year
   f. 0.033 million square kilometer per year
3. Even though the above data is accurate, give and explain two reasons why this dataset might lead you to an incorrect conclusion regarding global climate change.

Part B – South Pole Sea Ice Extent

Below is an expanded dataset showing Antarctic sea ice (Figure 6.7) extent measured during March 1980 through 2012 again downloaded from NSIDC. Only the even numbered years are presented, but the addition of odd years does not alter the trend in the data. Following the instructions in Figure 6.6, graph the data, draw a line of best fit, and calculate the slope of the line.

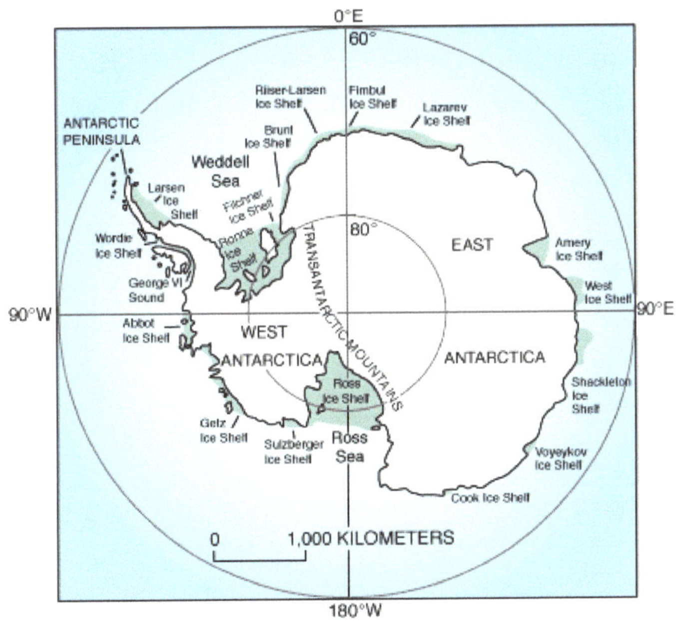

Figure 6.7 | Map of Antarctica showing the extent of the polar ice cap and the extent of the floating ice shelves.

year	Ice extent In million square km
1980	3.54
1982	4.8
1984	3.93
1986	4.08
1988	4.35
1990	4.36
1992	4.07
1994	5.07
1996	4.61
1998	4.65
2000	4.43
2002	3.94
2004	4.81
2006	3.44
2008	5.74
2010	4.03
2012	5.00

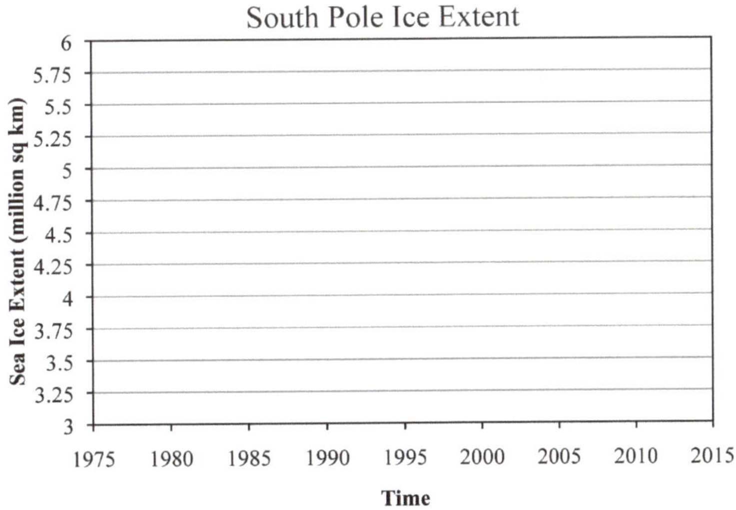

4. What is the slope of the line of best fit you estimated for this data set? Make sure to show your work.
5. What conclusion about climate change could you make from this dataset? How does your result for the extended dataset compare to the results from the data presented in the article (Part A)?

Part C – North Pole Sea Ice Extent

Next, we will examine the ice extent patterns of the northern Arctic polar ice sheet that is located around Greenland (Figure 6.8). The ice extent data is from March 1980 through 2012, for even numbered years, again downloaded from NSIDC. Following the instructions in Figure 6.6, graph the data, draw a line of best fit, and calculate the slope of the line.

Figure 6.8 | Map of Greenland showing the extent of the polar ice cap.

year	Ice extent in million square km
1980	16.13
1982	16.15
1984	15.62
1986	16.08
1988	16.13
1990	15.88
1992	15.47
1994	15.58
1996	15.13
1998	15.66
2000	15.27
2002	15.44
2004	15.05
2006	14.43
2008	15.22
2010	15.11
2012	15.21

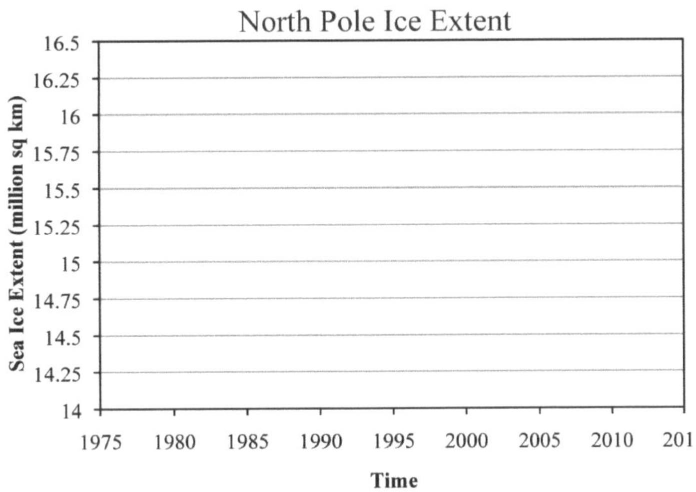

6. What is the slope of the line of best fit you estimated for this data set? Make sure to show your work.
7. What conclusion on climate change could you make from this dataset? How does your result for the North Pole compare to that of the South Pole (Part B)?

6.4 HEAT TRANSPORT AND OCEANIC CURRENTS

As previously stated in the laboratory, the tropics exhibit higher temperatures than the poles due to variations in the angle of solar radiation incidence on the Earth (Figure 6.1). Minimal solar radiation penetrates higher latitude regions due to its almost parallel incidence on the Earth's surface. Consequently, the majority of thermal energy at elevated latitudes is derived from the transfer of heat originating in the tropics. Heat is conveyed across the Earth's surface by atmospheric currents, meteorological phenomena, and oceanic currents. Large circular ocean currents, known as gyres, significantly influence the spatial distribution of heat on Earth and have played a crucial role in large-scale climate change throughout Earth's history. These currents are especially proficient at melting polar ice by affecting the sea ice from beneath. Furthermore, an analysis of ocean current patterns will elucidate the sea ice extent patterns previously graphed in relation to global warming (Parts B and C).

6.4.1 LAB EXERCISE

Materials

A visualization of the ocean currents can be seen by downloading the file "ocean_currents.kml" either from your course's website or directly from the Science on a Sphere page from NOAA (sos.noaa.gov/kml/). Once you download the file you can open it within Google Earth. Once the file loads (which may take a few minutes), click on the wrench in the upper left corner of the screen to the left of the NOAA logo (Figure 6.9). Check the "loop animation" box and slide the cursor for animation speed to an intermediate position, then click OK. Then click the toggle animation icon (Figure 6.9) to start the animation. Watch the movement of the currents as they flow, making sure to examine the flow in different parts of the world and zoom out to get a broad prospective of the flow of the ocean currents around the world.

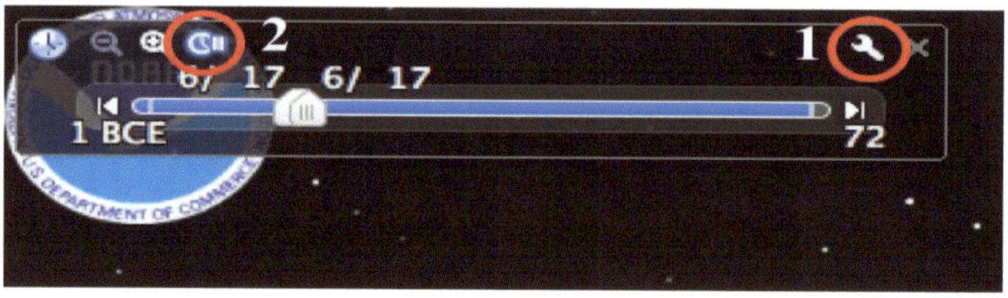

Figure 6.9 | Directions to visualize ocean currents in Google Earth. 1: Wrench icon to apply the correct settings. 2: Toggle animation icon to start the animation.

If you have difficulty loading the file you can also access the visualization at: http://sos.noaa.gov/Datasets/dataset.php?id=130#. To access the visualization of the ocean currents click on "Interactive Sphere" and hit play, make sure to spin the globe to see the currents in the different oceans.

Part D – Ocean Currents and Heat Transport

8. Examine and describe the ocean currents flowing in the Atlantic Ocean from the equator to the North Pole starting from Brazil (6 08 54.55S 35 58 24.09W). Do these currents reach the northern polar ice sheet? If the tropics become warmer, how would this affect the northern ice sheet?
9. Examine and describe the ocean currents flowing in the Pacific Ocean from the equator to the South Pole starting from Somalia (0 32 32.23N 44 09 15.47E). Do these currents reach the southern polar ice sheet? If the tropics become warmer, how would this affect the southern ice sheet?
10. Based on your answer to questions eight and nine, explain why we see different trends in the sea ice extent in the south (Part B) and north poles (Part C).
11. How might the changes you saw in the previous exercises relate to global albedo, sea level, ocean salinity, and temperature?

Part E – Concise Summary

Drawing conclusions, much less formulating policy decisions, about any intricate system like climate and its transformations is challenging. Accurate data can be distorted, and accurate data presented out of context (such as ice extent without consideration of ocean currents) may result in erroneous conclusions. Conclusions should be founded on rational, accurate, comprehensive, and contextual data, rather than on badly obtained or distorted information. It is crucial to draw your own conclusions from facts rather than being influenced by the author's perspective, applicable in both scientific contexts and various facets of life.

12. Do you think any conclusions (if any) regarding climate change should be made based on the data presented in this assignment?
13. What data do you think are needed to make a conclusion regarding climate change?

6.5 ADDITIONAL RESOURCES

To learn more about the science behind climate change and the scientific communities position on anthropogenic climate change see the following:

1. National Aeronautics and Space Administration (NASA) climate Site: http://climate.nasa.gov/
2. National Oceanic and Atmospheric Administration Climate Site: http://www.noaa.gov/climate.html
3. Geological Society of America's Position Statement on Climate Change: https://www.geosociety.org/gsa/positions/position10.aspx
4. American Metrological Society's Position Statement on Climate Change: https://www.ametsoc.org/ams/index.cfm/about-ams/ams-statements/state-ments-of-the-ams-in-force/climate-change/
5. American Chemical Society's Position Statement on Climate Change: https://www.acs.org/content/acs/en/policy/publicpolicies/sustainability/globalclimatechange.html?_ga=2.227911517.1368062825.1497897808 - 1921393895.1497558864
6. American Physical Society's Position Statement on Climate Change: http://www.aps.org/policy/statements/climate/
7. American Association for the Advancement of Science's Position Statement on Climate Change: http://www.aaas.org/sites/default/files/migrate/uploads/aaas_climate_ statement.pdf

6.6 STUDENT RESPONSES

1. Based exclusively on the data provided and your graph, what conclusion would you make regarding climate change?
   a. Sea ice is expanding, which indicates an increase in temperature
   b. Sea ice is expanding, which indicates a decrease in temperature
   c. Sea ice is contracting, which indicates an increase in temperature
   d. Sea ice is contracting, which indicates a decrease in temperature

2. What is slope of the line of best fit for this data?
   a. 0.008 million square kilometers per year
   b. 0.05 million square kilometers per year
   c. -0.05 million square kilometers per year
   d. 0.017 million square kilometers per year
   e. -0.17 million square kilometer per year
   f. 0.033 million square kilometer per year
3. Even though the above data is accurate, give and explain two reasons why this dataset might lead you to an incorrect conclusion regarding global climate change.
4. What is the slope of the line of best fit you estimated for this data set? Make sure to show your work.
5. What conclusion about climate change could you make from this dataset? How does your result for the extended dataset compare to the results from the data presented in the article (Part A)?
6. What is the slope of the line of best fit you estimated for this data set? Make sure to show your work.
7. What conclusion on climate change could you make from this dataset? How does your result for the North Pole compare to that of the South Pole (Part B)?
8. Examine and describe the ocean currents flowing in the Atlantic Ocean from the equator to the North Pole starting from Brazil (6 08 54.55S 35 58 24.09W). Do these currents reach the northern polar ice sheet? If the tropics become warmer, how would this affect the northern ice sheet?
9. Examine and describe the ocean currents flowing in the Pacific Ocean from the equator to the South Pole starting from Somalia (0 32 32.23N 44 09 15.47E). Do these currents reach the southern polar ice sheet? If the tropics become warmer, how would this affect the southern ice sheet?
10. Based on your answer to questions eight and nine, explain why we see different trends in the sea ice extent in the south (Part B) and north poles (Part C).
11. How might the changes you saw in the previous exercises relate to global albedo, sea level, ocean salinity, and temperature?
12. Do you think any conclusions (if any) regarding climate change should be made based on the data presented in this assignment?
13. What data do you think are needed to make a conclusion regarding climate change?

7 Chapter | MATTER AND MINERALS

7.1 INTRODUCTION: AN OVERVIEW

Have you utilized a mineral today? Although numerous individuals may initially respond negatively, consider the following inquiries: Have you performed oral hygiene by brushing your teeth? Have you consumed any food that may include salt? Did you use make-up this morning, or do you have polished fingernails or toenails? Have you utilized a mobile phone? What about utilizing a car, bicycle, or public transit? If you have engaged in any of those activities, you have utilized at least one mineral, and in numerous instances, you have employed a considerable quantity of minerals. Minerals are prevalent and beneficial in everyday products, although most individuals remain unaware of this fact.

A mineral is characterized as a naturally occurring, inorganic solid possessing a certain chemical composition and a distinct crystalline structure. Let us deconstruct that definition.

Figure 7.1 | Synthetic bismuth

By "naturally occurring," it signifies that any substance made by humans, such as the aesthetically pleasing synthetic bismuth depicted in Figure 7.1, is excluded from the classification of a mineral. A mineral qualifies as an inorganic solid if it lacks complex carbon molecules associated with life and exists in a solid state, as opposed to vapor or liquid. This indicates that water, in its liquid form, is not classified as a mineral, yet ice, in its solid state, qualifies as a mineral, provided it is not artificially produced. A specific chemical composition denotes the chemical formula of a mineral. For the majority of minerals, this remains constant (e.g., halite is NaCl); however, certain minerals exhibit compositional variability, since one element may substitute for another of comparable size and charge (e.g., olivine is (Mg,Fe)2SiO4, with variable magnesium and iron content). The atoms in minerals are arranged systematically, making the distinctive crystalline structure a reflection of the internal atomic configuration.

Minerals are significant not just for their diverse applications but also as the fundamental components of rocks. This laboratory session will establish the groundwork for all subsequent rock laboratories in the course. Accurate mineral identification is essential in geology; thus, proceed through this laboratory exercise meticulously. Numerous minerals exist, although we shall concentrate on merely eighteen of the most prevalent.

7.2 PHYSICAL PROPERTIES

Identifying a mineral resembles the process of detective work. Minerals are characterized by their physical qualities. Refer to Figure 7.2, for instance. What is your description of it? It can be described as glossy, gold, and has a distinct shape. Each of these descriptions corresponds to a physical attribute (shiny=luster, gold=color, shape=crystal form). Physical characteristics may differ among identical minerals, necessitating vigilance. Color is a characteristic that often lacks reliability as a diagnostic instrument in numerous instances. Quartz is a mineral that exhibits a range of hues, as illustrated in Figure 7.3. At times, color can be advantageous, as exemplified by the mineral olivine. Olivine is described as "olive green," encompassing a range from pale to dark green, as illustrated in Figure 7.4. Exercise caution when utilizing color to detect minerals. We will examine each physical property in depth to assist you in identifying the minerals.

Figure 7.2 | Describe this mineral.

Figure 7.3 | Examples of the different varieties of quartz (jasper, rose quartz, smoky quartz, agate, amethyst, citrine, and petrified wood), demonstrating the difficulty of identifying this mineral.

Figure 7.4 | The mineral olivine is "olive green."

7.2.1 Hardness

Hardness denotes a mineral's resistance to being scratched by another mineral or material, resulting from the strength of the atomic bonds inside the mineral. The substance causing the scratching is harder than the item being scratched. Hardness is determined by a scale ranging from 1 to 10, established by the mineralogist Friedrich Mohs (Figure 7.5). Mohs' scale enumerates eleven minerals arranged by their relative hardness. Each mineral on the scale can abrade a mineral with a lower numerical value. Your mineral kit includes various products of established hardness. The glass plate possesses a hardness of 5.5, the iron nail has a hardness of 4, the copper wire exhibits a hardness of 3, and the fingernail demonstrates a hardness of 2.5. If a mineral can be scratched, it is softer than a fingernail, indicating its hardness is less than 2.5. Apply force when attempting to scrape a surface, but exercise caution with the glass plate. Consistently place the glass plate on a level surface instead than holding it in your hand to prevent breakage. Do not conflate mineral powder with a scratch; utilize your finger to detect a groove formed by a scratch (mineral powder is deposited when a softer mineral abrades a harder surface). Materials of comparable hardness struggle to scratch one another; for instance, your fingernail may not consistently be able to scratch biotite mica, which possesses a hardness of 2.5..

Number	Mineral	Hardness of Test Kit Items
1	Talc	(softest mineral)
2	Gypsum	2.5 – Fingernail
3	Calcite	3 – Copper Wire
4	Fluorite	4 – Nail
5	Apatite	5.5 – Glass Plate
6	Orthoclase Feldspar	
7	Quartz	
8	Topaz	
9	Corundum	
10	Diamond	(hardest mineral)

Figure 7.5 | Mohs' Scale of Hardness

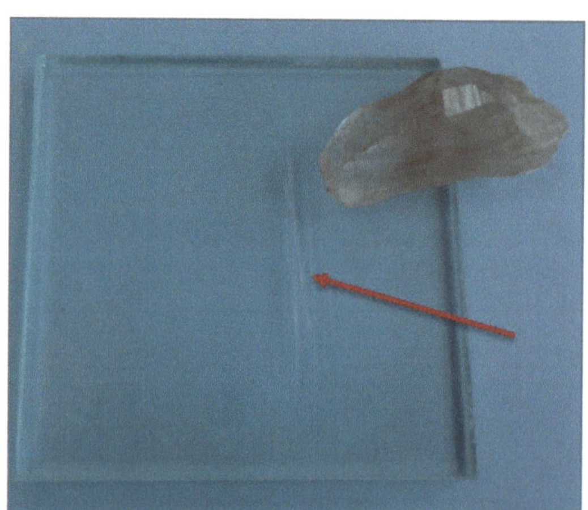

Figure 7.6 | An example of a scratch made by the mineral quartz on a streak plate. The red arrow is point ing to the scratch. Quartz, therefore, is harder than glass.

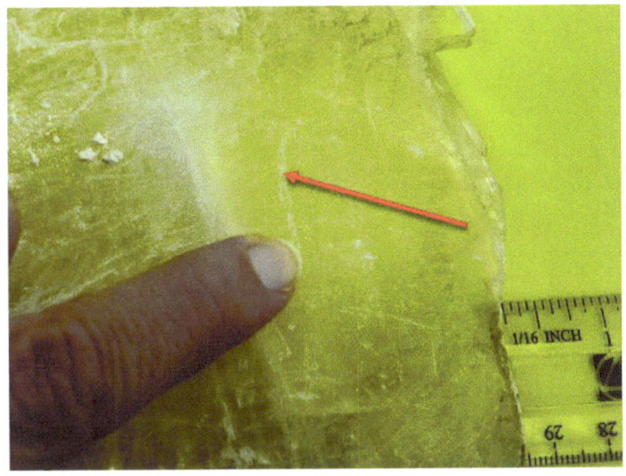

Figure 7.7 | An example of a scratch made by a fingernail on the mineral gypsum. The red arrow is pointing to the scratch. Gypsum, therefore, is softer than a fingernail.

7.2.2 LAB EXERCISE

Materials

Your HOL Lab Kit contains 18 numbered mineral samples, separated into 3 bags (labeled as Mineral Bag 1, 2, or 3). Use these instructions to test and identify them. You will test for different properties after learning about them, then work on identification at the end of the lab. The HOL kit has been specifically tailored to this class – make sure that you are using the kit required by this class, as other rock and mineral kits will not work. Images will be provided of the correct kit; make sure that you closely compare your kit to the images so that you are working with the correct samples. Empty the contents of the testing kit. It will contain:

a. A 3" copper wire
b. Glass plate (wrapped in paper) – this will be used in testing hardness
c. Zinc coated nail
d. Unglazed porcelain plate (wrapped in paper) – this will be used as a streak plate
e. Hydrochloric acid
f. Magnifying glass (10x). To use this, hold it very close to your eye and bring the sample near the glass until it is in focus (approximately one inch from your eye).
g. Gloves and protective goggles (for use with the acid)

Take out Minerals Bag 1 and lay the six mineral samples out on a white sheet of paper. It should appear like Figure 7.8. We will first examine hardness from these six samples, and will answer more questions about them later in the lab. Look closely at each of the minerals, using the hand lens to observe them. In this bag, you have the following minerals (not listed in order): Microcline (also called Potassium Feldspar), Fluorite, Quartz, Olivine, Talc, and Selenite (also called Gypsum). They are numbered 1-6.

You need to experiment with each sample to test for its hardness and use Figure 7.5 for reference. Remember that hardness is determined by scratching the mineral (or using the mineral to scratch something else). First, decide which minerals have a hardness greater than 5.5 (the hardness of glass). Lay the glass on a flat surface (not in your hand), then try to scratch it with each mineral. Bare down hard with the mineral, much like trying to leave a scratch on a car with a key. Table 7.1 is given at the end of this lab for you to make notations about each mineral. Note that you do not have to fill in every physical property for every mineral (that would be very time-consuming with 18 samples). Just fill in the properties you are asked about as you work. Note now on the table which minerals have a hardness greater than 5.5. You may also test samples by using materials to scratch them. The copper wire has a hardness of 3. Any mineral that it can scratch will have a hardness less than 3. You can further refine this by using your fingernail (only natural fingernails work for this). Your fingernail has a hardness of 2.5, so if the copper wire scratches a mineral and your fingernail also scratches it, you know its hardness must be <2.5. The zinc coated nail has a hardness of 4. Also use it to scratch the minerals. Minerals may also be used to scratch each other. For example, if you have two minerals that have a hardness of <2.5, you can see if one will scratch the other. Then you know it is harder, since it did the scratching.

Figure 7.8 | The six minerals (#1-6) in Minerals Bag 1 in the HOL kit.

1. Sample 1: What is this sample's hardness?
   a. harder than glass
   b. softer than glass but harder than nail
   c. softer than nail but harder than copper
   d. softer than copper but harder than a fingernail
   e. softer than a fingernail

2. Sample 2: What is this sample's hardness?
   a. harder than glass
   b. softer than glass but harder than nail
   c. softer than nail but harder than copper
   d. softer than copper but harder than a fingernail
   e. softer than a fingernail
3. Sample 3: What is this sample's hardness?
   a. harder than glass
   b. softer than glass but harder than nail
   c. softer than nail but harder than copper
   d. softer than copper but harder than a fingernail
   e. softer than a fingernail
4. Sample 4: What is this sample's hardness?
   a. harder than glass
   b. softer than glass but harder than nail
   c. softer than nail but harder than copper
   d. softer than copper but harder than a fingernail
   e. softer than a fingernail
5. Sample 5: What is this sample's hardness?
   a. harder than glass
   b. softer than glass but harder than nail
   c. softer than nail but harder than copper
   d. softer than copper but harder than a fingernail
   e. softer than a fingernail
6. Sample 6. What is this sample's hardness?
   a. harder than glass
   b. softer than glass but harder than nail
   c. softer than nail but harder than copper
   d. softer than copper but harder than a fingernail
   e. softer than a fingernail

7.3 CRYSTAL FORM

This feature pertains to the geometric form that a crystal inherently assumes, reflecting the systematic internal arrangement of atoms within the mineral.

Minerals will exhibit their crystalline structure if they have sufficient space for growth during their development. Optimal growth circumstances are rarely consistently present; hence, numerous minerals fail to exhibit their optimal crystal form due to overcrowded situations during their development. Figure 7.9 illustrates examples of crystal morphology.

Cube Hexagonal Prism Rhombohedron

Octahedron (8 faces) Dodecahedron (12 faces)

Figure 7.9 | Examples of crystal form

7.4 CLEAVAGE

When minerals are fractured, such as with a rock hammer, some may cleave along smooth, flat surfaces referred to as cleavage planes. The flat surfaces are aligned with the crystal's directions of weakness. The links between atoms in a mineral may vary in strength, resulting in fractures occurring along these zones of weakness when the crystal is broken. This produces planar cleavage surfaces. Minerals exhibiting excellent cleavage fracture in a smooth, flat plane, whereas those with poor cleavage fracture in a more uneven manner. Certain minerals lack zones of weakness due to uniform bond strength or misalignment of weaker bonds within a plane. If this is the case, it will not exhibit cleavage but will instead fracture in a random and uneven manner. Ensure the differentiation between cleavage and crystal form. Crystal shape develops during the growth of a mineral, whereas cleavage manifests when a mineral fractures. Refer to Figure 7.10 for the primary forms of cleavage and corresponding examples.

# of Cleavages & Direction	Cleavage Name	Example
0 (none) – mineral fractures	No cleavage planes	
1	Basal cleavage – flat sheets	
2 – cleavages at or near 90°	Prismatic cleavage – rectangular cross-sections	
2 – cleavages not at 90°	Prismatic cleavage – parallelogram cross-sections	

3 – cleavages at 90o	Cubic cleavage – cubes	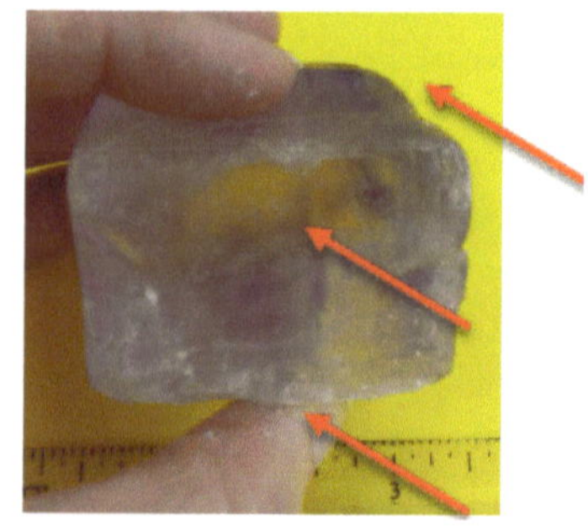
3 – cleavages not at 90o	Rhombohedral cleavage – rhombs	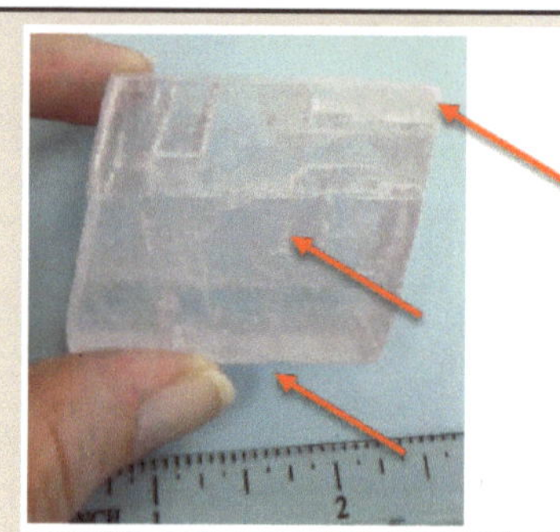
4	Octahedral cleavage	

Figure 7.10 | Chart with the main types of cleavage, along with picture examples. Red arrows are pointing to the different cleavage planes in each picture.

A mineral may have one or more cleavage planes. Planes that are parallel are considered the same direction of cleavage and should only count as one. One direction of cleavage is termed basal cleavage. Minerals that display this cleavage will break off in flat sheets. Two directions of cleavage is termed prismatic, while three directions of cleavage at 90o is referred to as cubic. A mineral with four directions of cleavage is termed octahedral. With 2 or more cleavage planes present, it is important to pay attention to the angle of the cleavage planes. To determine the angle of cleavage, look at the intersection of cleavage planes. Commonly, cleavage planes will intersect at 60o, 90o (right angles), or 120o. Be cautious when you see a flat surface on a mineral – not every flat surface is a cleavage plane. Crystal faces can be flat, but remember they form as a mineral grows, while cleavage forms as a mineral breaks. The crystal form of quartz is a hexagonal prism, with nice flat sides. But when quartz is hit with a rock hammer, it breaks in an irregular fashion and does not exhibit cleavage. Also use caution when trying to distinguish the minerals pyroxene and amphibole. Both minerals are black or greenish-black, with similar hardness, making them difficult to tell apart. You must observe the cleavage angles to tell them apart. Cleavage angles in pyroxene are near 90o, so expect it to look boxy and form right angles. Cleavage angles in amphibole are 60o and 120o, so expect a more bladed or pyramid like appearance (Figure 7.11).

Amphibole – cleavage angles at 60° and 120°	**Pyroxene** – cleavage angles near 90°

Figure 7.11 | Comparison of cleavage angles between amphibole and pyroxene.

7.5 FRACTURE

When minerals fail to cleave along designated planes and instead exhibit erratic breakage, they are described as fracturing. Fracture surfaces are typically either irregular or conchoidal, exhibiting a ribbed, smoothly curved texture akin to shattered glass. (Figure 7.12).

Figure 7.12 | This piece of igneous rock called obsidian has been hit with a hammer and is displaying conchoidal fracture.

7.5.1 LAB EXERCISE

Take out Minerals Bag 2 and lay the six mineral samples out on a white sheet of paper. It should appear like Figure 7.13. We will first examine cleavage and fracture, along with hardness, from these six samples, and will answer more questions about them later in the lab. Look closely at each of the minerals, using the hand lens to observe them. In this bag, you have the following minerals (not listed in order): Pyroxene, Muscovite Mica, Halite, Amphibole, Calcite, and Biotite Mica. They are numbered 7-12.

Figure 7.13 | The six minerals (#7-12) in Minerals Bag 2 in the HOL kit.

7. Sample 7: This sample has:
   a. no cleavage (it fractures)	b. 1 cleavage plane
   c. 2 cleavage planes at 90o	d. 3 cleavage planes at 90o
   e. 4 cleavage planes
8. Sample 8: This sample has:
   a. no cleavage (it fractures)	b. 1 cleavage plane
   c. 2 cleavage planes not at 90o	d. 3 cleavage planes not at 90o
   e. 4 cleavage planes
9. Sample 9: This sample has:
   a. no cleavage (it fractures)	b. 1 cleavage plane
   c. 2 cleavage planes not at 90o	d. 3 cleavage planes at 90o
   e. 4 cleavage planes

10. Sample 10: This sample has:
    a. no cleavage (it fractures)
    b. 1 cleavage plane
    c. 2 cleavage planes not at 90o
    d. 2 cleavage planes at 90o
    e. 4 cleavage planes
11. Sample 11: This sample has:
    a. no cleavage (it fractures)
    b. 1 cleavage plane
    c. 2 cleavage planes not at 90o
    d. 2 cleavage planes at 90o
    e. 3 cleavage planes at 90o
12. Sample 11: What is this sample's hardness?
    a. harder than glass
    b. softer than glass but harder than nail
    c. softer than nail but harder than copper
    d. softer than copper but harder than a fingernail
    e. softer than a fingernail
13. Sample 12: This sample has:
    a. no cleavage (it fractures)
    b. 1 cleavage plane
    c. 2 cleavage planes not at 90o
    d. 2 cleavage planes at 90o
    e. 3 cleavage planes not at 90o
14. Sample 12: What is this sample's hardness?
    a. harder than glass
    b. softer than glass but harder than nail
    c. softer than nail but harder than copperd.
    d. softer than copper but harder than a fingernail
    e. softer than a fingernail

7.6 LUSTER

Luster denotes the visual quality of light reflection from a mineral's surface. It is often categorized into two primary types: metallic and non-metallic. Minerals exhibiting a metallic shine possess the coloration characteristic of metals such as silver, gold, copper, or brass (Figure 7.14). Although minerals exhibiting a metallic shine are frequently lustrous, not all lustrous minerals possess metallic characteristics. Ensure you observe the hue of the metal, rather than merely its luster. Minerals exhibiting non-metallic sheen do not resemble metals. They may exhibit vitreous (glassy), earthy (dull), waxy (resembling a candle's shine), greasy (oily), or various other lusters. (Figure 7.15).

Figure 7.14 | Examples of the metallic luster of pyrite, also known as "fool's gold."

Vitreous	
Earthy	
Waxy	

Figure 7.15 | Examples of different types of non-metallic lusters.

7.7 STREAK

Streak is an easily detectable physical property. It refers to the color left behind on an unglazed piece of porcelain when a mineral is rubbed along its surface. A streak plate is included in your rock and mineral kit to test this property. Often a mineral will have a streak of a different color than the color of the mineral (for example, pyrite has a dark gray streak, Figure 7.16). Some minerals will have a white streak, which is difficult to see along the white streak plate. If you rub a mineral along the streak plate and do not see an obvious streak, wipe your finger along the streak plate. A mineral with a white streak will leave a white powder behind that will rub on your finger (Figure 7.17).

Figure 7.16 | An example of the dark gray streak left behind when pyrite is rubbed along a streak plate.

Figure 7.17 | An example of the white streak (on finger left behind when flurite is rubbed along a streak plate.

7.8 SPECIAL PHYSICAL PROPERTIES

Several minerals have unique properties that aid in their identification. **Tenacity** refers to the way a mineral resists breakage. If a mineral shatters like glass, it is said to be brittle (like quartz), while minerals that can be hammered are malleable (like copper, Figure 7.18). Minerals may be elastic, in which they are flexible and bend like a plastic comb, but return to their original shape (like mica, Figure 7.19). Sectile minerals are soft like wax, and can be separated with a knife (like gypsum).

Figure 7.18 | Copper, which can be hammered into thin sheets, is malleable.

Some minerals react when dilute hydrochloric acid is placed on them. Carbonate minerals (minerals that include CO in their chemical formula) will effect or fizz when acid is applied to them. When you test a mineral with acid, be cautious and use just a drop of the acid. Use your magnifying glass to look closely for bubbles (Figure 7.20). The acid is very dilute and will not burn your skin or clothing, but wash your hands after use (gloves and goggles are provided). Also make sure that you rinse with water and wipe off the acid from the minerals that you test.

Minerals may be magnetic, and this property is simply tested by seeing if your nail is attracted to a mineral. Magnetite is an example of a magnetic mineral. The mineral halite is simply table salt, so it will taste salty. Graphite is used in pencils, and makes a nice smudge when rubbed along paper. Talc will feel soapy when touched.

Figure 7.19 | Muscovite mica, which bends but returns to its original shape, is elastic.

Specific gravity is the ratio of a mineral's weight to the weight of an equal volume of water. A mineral with a specific gravity of 2 would weigh twice as much as water. Most minerals are heavier than water, and the average specific gravity for all minerals is approximately 2.7. Some minerals are quite heavy, such as pyrite with a specific gravity of 4.9-5.2, native copper, with a specific gravity of 8.8-9.0, and native gold at 19.3, which makes panning useful for gold, as the heavy mineral stays behind as you wash material out of the pan.

Figure 7.20 | Note the effervescing acid bubbles at the red arrow on this piece of calcite.

7.8.1 LAB EXERCISE

Take out Minerals Bag 3 and lay the six mineral samples out on a white sheet of paper. It should appear like Figure 7.21. We will first examine several properties, including streak, from these six samples, and will answer more questions about them later in the lab. Look closely at each of the minerals, using the hand lens to observe them. In this bag, you have the following minerals (not listed in order): Magnetite, Graphite, Copper, Sulfur, Hematite, and Pyrite. They are numbered 13-18.

Figure 7.21 | The six minerals (#13-18) in Minerals Bag 3 in the HOL kit.

15. Sample 13: What is the streak of this sample?
    a. dark gray streak
    b. white streak
    c. reddish brown streak
    d. pale yellow streak
16. Sample 13: What is the luster of this sample?
    a. non-metallic, vitreous
    b. non-metallic, earthy
    c. non-metallic, greasy
    d. non-metallic, waxy
    e. metallic
17. Sample 14: What is the streak of this sample?
    a. dark gray to black streak
    b. white streak
    c. reddish brown streak
    d. pale yellow streak
18. Sample 14: Which other item(s) is/are characteristic(s) of this sample?
    a. stains the fingers
    b. harder than glass
    c. greasy feel
    d. both a and b
    e. both a and c
19. Sample 15: What is the streak of this sample?
    a. dark gray to black streak
    b. white streak
    c. reddish brown streak
    d. pale yellow streak
20. Sample 16: What is the luster of this sample?
    a. non-metallic, vitreous
    b. non-metallic, earthy
    c. non-metallic, greasy
    d. non-metallic, waxy
    e. metallic
21. Sample 16: Due to its appearance, this sample has often been confused with native gold, a mineral with a hardness of 2.5-3. How does its hardness compare with that of gold?
    a. Sample 16 is harder than gold.
    b. Sample 16 is softer than gold.
22. Sample 17: What is the streak of this sample?
    a. dark gray to black streak
    b. white streak
    c. reddish brown streak
    d. pale yellow streak
23. Sample 17: What another unique property does this sample have?
    a. effervescence in acid
    b. it is magnetic
    c. it tastes salty
    d. it feels soapy
    e. it writes on paper

24. Sample 18. Examine this entire sample closely with a hand lens. What is the luster of this sample?

    a. non-metallic, vitreous　　b. non-metallic, earthy

    c. non-metallic, greasy　　d. non-metallic, waxy

    e. metallic

Now that you have had practice at detecting the properties of your 18 mineral samples, take the next step of identifying each sample and answering the questions below. Use the Mineral Identification Chart (Figure 7.22) to help you.

25. Sample 1: What is this sample?

    a. Microcline　　b. Fluorite　　c. Quartz

    d. Olivine　　e. Talc　　f. Selenite (Gypsum)

26. Sample 2: What is this sample?

    a. Microcline　　b. Fluorite　　c. Quartz

    d. Olivine　　e. Talc　　f. Selenite (Gypsum)

Luster	Hardness	Cleavage	Other Properties	Mineral Name
Non-Metallic	> Glass	Poor Cleavage	Red-brown, black, silver in color. H=6. St=red-brown	Hematite
			Olive-green in color. H=6. St=white. Commonly granular	Olivine
			Variety of colors. H=7. Conchoidal fracture. Vitreous luster.	Quartz
		Clearly Shows Cleavage	Black to greenish black in color. H=6. C=2 planes at $\sim 60^{o}$ and 120°. Elongated crystals.	Amphibole
			Tan-pink, white, green in color. H=6. C=2 planes at 90°.	Microcline
			Black to greenish black in color. H=6. C=2 planes at $\sim 90^{o}$. Short stubby crystals.	Pyroxene
	< Glass	Poor Cleavage	Dark gray to black in color. H=1. Greasy feel will smudge fingers.	Graphite
			Yellow in color. H=1.5-2.5. St=white to yellow.	Sulfur
			White to green in color. H=1. Soapy feel.	Talc

			Brown to black in color. H=2.5. C=1 perfect. Breaks into thin sheets that are elastic.	Biotite Mica
			White to transparent in color. H=3. C=3 rhombohedral. Strong effervescence in acid.	Calcite
		Clearly Shows Cleavage	Transparent, yellow, purple, green in color. H=4. C=4 octahedral.	Fluorite
			Transparent to white in color. H=2. C=3, though 2 directions may be difficult to see.	Gypsum
			White to transparent in color. H=2.5. C=3, cubic. Tastes salty.	Halite
			Transparent, light brown, to yellow in color. H=2.5. C=1 perfect. Breaks into thin sheets that are elastic.	Muscovite Mica
Metallic	> Glass	Poor Cleavage	Black in color. H=6. St=black. Strongly magnetic.	Magnetite
			Brass-yellow in color. H=6.5. St=dark gray.	Pyrite
	<Glass	Poor Cleavage	Copper-red in color. Tarnishes to green or black in air. H=2.5-3. St=copper-red.	Copper

Figure 7.22 | Mineral Identification Chart

27. Sample 3: What is this sample?
    a. Microcline b. Fluorite c. Quartz
    d. Olivine e. Talc f. Selenite (Gypsum)
28. Sample 4: What is this sample?
    a. Microcline b. Fluorite c. Quartz
    d. Olivine e. Talc f. Selenite (Gypsum)
29. Sample 4: What other unique property does this sample have?
    a. effervescence in acid b. it is magnetic
    c. it tastes salty d. it feels soapy
    e. it writes on paper
30. Sample 5: What is this sample?
    a. Microcline b. Fluorite c. Quartz
    d. Olivine e. Talc f. Selenite (Gypsum)

31. Sample 6: What is this sample?
    a. Microcline b. Fluorite c. Quartz
    d. Olivine e. Talc f. Selenite (Gypsum)

32. Sample 7: What is this sample?
    a. Pyroxene b. Muscovite Mica c. Halite
    d. Amphibole e. Calcite f. Biotite Mica

33. Sample 7: What other unique property does this sample have?
    a. effervescence in acid b. it is magnetic
    c. it tastes salty d. it feels soapy
    e. it writes on paper

34. Sample 8: What is this sample?
    a. Pyroxene b. Muscovite Mica
    c. Halite d. Amphibole
    e. Calcite f. Biotite Mica

35. Sample 8: What other unique property does this sample have?
    a. effervescence in acid b. it is magnetic
    c. it tastes salty d. it feels soapy
    e. it writes on paper

36. Sample 9: What is this sample?
    a. Pyroxene b. Muscovite Mica
    c. Halite d. Amphibole
    e. Calcite f. Biotite Mica

37. Sample 10: What is this sample?
    a. Pyroxene b. Muscovite Mica
    c. Halite d. Amphibole
    e. Calcite f. Biotite Mica

38. Sample 10: Test this tenacity of this mineral by trying to bend it. Which way does it behave?
    a. sectile b. malleable c. elastic d. brittle

39. Sample 11: What is this sample?
    a. Pyroxene b. Muscovite Mica
    c. Halite d. Amphibole
    e. Calcite f. Biotite Mica

40. Sample 12: What is this sample?
    a. Pyroxene
    b. Muscovite Mica
    c. Halite
    d. Amphibole
    e. Calcite
    f. Biotite Mica
41. Sample 13: What is this sample?
    a. Magnetite
    b. Graphite
    c. Copper
    d. Sulfur
    e. Hematite
    f. Pyrite
42. Sample 14: What is this sample?
    a. Magnetite
    b. Graphite
    c. Copper
    d. Sulfur
    e. Hematite
    f. Pyrite
43. Sample 15: What is this sample?
    a. Magnetite
    b. Graphite
    c. Copper
    d. Sulfur
    e. Hematite
    f. Pyrite
44. Sample 16: What is this sample?
    a. Magnetite
    b. Graphite
    c. Copper
    d. Sulfur
    e. Hematite
    f. Pyrite
45. Sample 17: What is this sample?
    a. Magnetite
    b. Graphite
    c. Copper
    d. Sulfur
    e. Hematite
    f. Pyrite
46. Sample 18: What is this sample?
    a. Magnetite
    b. Graphite
    c. Copper
    d. Sulfur
    e. Hematite
    f. Pyrite

Table 7.1 | Mineral Notation Chart – Fill in this chart as you work through the lab. An example of a mineral you do not have in your kit (#0) is included. You do not have to fill out every column for every mineral – just follow along in the lab and determine the properties you are asked about.

Mineral #	Luster	Hardness	Cleavage/ Fracture	Streak	Other Notable Properties (include color when diagnostic)	Name
0	Metallic	~2.5 – may scratch fingernail	3 cubic	gray	High specific gravity because it is heavy	Galena

7.9 STUDENT RESPONSES

The following is a summary of the questions in this lab for ease in submitting answers online.

1. Sample 1: What is this sample's hardness?
   a. harder than glass
   b. softer than glass but harder than nail
   c. softer than nail but harder than copper
   d. softer than copper but harder than a fingernail
   e. softer than a fingernail

2. Sample 2: What is this sample's hardness?
   a. harder than glass
   b. softer than glass but harder than nail
   c. softer than nail but harder than copper
   d. softer than copper but harder than a fingernail
   e. softer than a fingernail
3. Sample 3: What is this sample's hardness?
   a. harder than glass
   b. softer than glass but harder than nail
   c. softer than nail but harder than copper
   d. softer than copper but harder than a fingernail
   e. softer than a fingernail
4. Sample 4: What is this sample's hardness?
   a. harder than glass
   b. softer than glass but harder than nail
   c. softer than nail but harder than copper
   d. softer than copper but harder than a fingernail
   e. softer than a fingernail
5. Sample 5: What is this sample's hardness?
   a. harder than glass
   b. softer than glass but harder than nail
   c. softer than nail but harder than copper
   d. softer than copper but harder than a fingernail
   e. softer than a fingernail
6. Sample 6. What is this sample's hardness?
   a. harder than glass
   b. softer than glass but harder than nail
   c. softer than nail but harder than copper
   d. softer than copper but harder than a fingernail
   e. softer than a fingernail
7. Sample 7: This sample has:
   a. no cleavage (it fractures)
   b. 1 cleavage plane
   c. 2 cleavage planes at 90o
   d. 3 cleavage planes at 90o
   e. 4 cleavage planes

8. Sample 8: This sample has:
   a. no cleavage (it fractures)
   b. 1 cleavage plane
   c. 2 cleavage planes not at 90o
   d. 3 cleavage planes not at 90o
   e. 4 cleavage planes
9. Sample 9: This sample has:
   a. no cleavage (it fractures)
   b. 1 cleavage plane
   c. 2 cleavage planes not at 90o
   d. 3 cleavage planes at 90o
   e. 4 cleavage planes
10. Sample 10: This sample has:
    a. no cleavage (it fractures)
    b. 1 cleavage plane
    c. 2 cleavage planes not at 90o
    d. 2 cleavage planes at 90o
    e. 4 cleavage planes
11. Sample 11: This sample has:
    a. no cleavage (it fractures)
    b. 1 cleavage plane
    c. 2 cleavage planes not at 90o
    d. 2 cleavage planes at 90o
    e. 3 cleavage planes at 90o
12. Sample 11: What is this sample's hardness?
    a. harder than glass
    b. softer than glass but harder than nail
    c. softer than nail but harder than copper
    d. softer than copper but harder than a fingernail
    e. softer than a fingernail
13. Sample 12: This sample has:
    a. no cleavage (it fractures)
    b. 1 cleavage plane
    c. 2 cleavage planes not at 90o
    d. 2 cleavage planes at 90o
    e. 3 cleavage planes not at 90o
14. Sample 12: What is this sample's hardness?
    a. harder than glass
    b. softer than glass but harder than nail
    c. softer than nail but harder than copper
    d. softer than copper but harder than a fingernail
    e. softer than a fingernail

15. Sample 13: What is the streak of this sample?
    a. dark gray streak
    b. white streak
    c. reddish brown streak
    d. pale yellow streak
16. Sample 13: What is the luster of this sample?
    a. non-metallic, vitreous
    b. non-metallic, earthy
    c. non-metallic, greasy
    d. non-metallic, waxy
    e. metallic
17. Sample 14: What is the streak of this sample?
    a. dark gray to black streak
    b. white streak
    c. reddish brown streak
    d. pale yellow streak
18. Sample 14: Which other item(s) is/are characteristic(s) of this sample?
    a. stains the fingers
    b. harder than glass
    c. greasy feel
    d. both a and b
    e. both a and c
19. Sample 15: What is the streak of this sample?
    a. dark gray to black streak
    b. white streak
    c. reddish brown streak
    d. pale yellow streak
20. Sample 16: What is the luster of this sample?
    a. non-metallic, vitreous
    b. non-metallic, earthy
    c. non-metallic, greasy
    d. non-metallic, waxy
    e. metallic
21. Sample 16: Due to its appearance, this sample has often been confused with native gold, a mineral with a hardness of 2.5-3. How does its hardness compare with that of gold?
    a. Sample 16 is harder than gold.
    b. Sample 16 is softer than gold.
22. Sample 17: What is the streak of this sample?
    a. dark gray to black streak
    b. white streak
    c. reddish brown streak
    d. pale yellow streak
23. Sample 17: What another unique property does this sample have?
    a. effervescence in acid
    b. it is magnetic
    c. it tastes salty
    d. it feels soapy
    e. it writes on paper

24. Sample 18. Examine this entire sample closely with a hand lens. What is the luster of this sample?
    a. non-metallic, vitreous
    b. non-metallic, earthy
    c. non-metallic, greasy
    d. non-metallic, waxy
    e. metallic
25. Sample 1: What is this sample?
    a. Microcline
    b. Fluorite
    c. Quartz
    d. Olivine
    e. Talc
    f. Selenite (Gypsum)
26. Sample 2: What is this sample?
    a. Microcline
    b. Fluorite
    c. Quartz
    d. Olivine
    e. Talc
    f. Selenite (Gypsum)
27. Sample 3: What is this sample?
    a. Microcline
    b. Fluorite
    c. Quartz
    d. Olivine
    e. Talc
    f. Selenite (Gypsum)
28. Sample 4: What is this sample?
    a. Microcline
    b. Fluorite
    c. Quartz
    d. Olivine
    e. Talc
    f. Selenite (Gypsum)
29. Sample 4: What other unique property does this sample have?
    a. effervescence in acid
    b. it is magnetic
    c. it tastes salty
    d. it feels soapy
    e. it writes on paper
30. Sample 5: What is this sample?
    a. Microcline
    b. Fluorite
    c. Quartz
    d. Olivine
    e. Talc
    f. Selenite (Gypsum)
31. Sample 6: What is this sample?
    a. Microcline
    b. Fluorite
    c. Quartz
    d. Olivine
    e. Talc
    f. Selenite (Gypsum)

32. Sample 7: What is this sample?
    a. Pyroxene b. Muscovite Mica
    c. Halite d. Amphibole
    e. Calcite f. Biotite Mica

33. Sample 7: What other unique property does this sample have?
    a. effervescence in acid b. it is magnetic
    c. it tastes salty d. it feels soapy
    e. it writes on paper

34. Sample 8: What is this sample?
    a. Pyroxene b. Muscovite Mica
    c. Halite d. Amphibole
    e. Calcite f. Biotite Mica

35. Sample 8: What other unique property does this sample have?
    a. effervescence in acid b. it is magnetic
    c. it tastes salty d. it feels soapy
    e. it writes on paper

36. Sample 9: What is this sample?
    a. Pyroxene b. Muscovite Mica
    c. Halite d. Amphibole
    e. Calcite f. Biotite Mica

37. Sample 10: What is this sample?
    a. Pyroxene b. Muscovite Mica
    c. Halite d. Amphibole
    e. Calcite f. Biotite Mica

38. Sample 10: Test this tenacity of this mineral by trying to bend it. Which way does it behave?
    a. sectile b. malleable c. elastic d. brittle

39. Sample 11: What is this sample?
    a. Pyroxene b. Muscovite Mica
    c. Halite d. Amphibole
    e. Calcite f. Biotite Mica

40. Sample 12: What is this sample?
    a. Pyroxene
    b. Muscovite Mica
    c. Halite
    d. Amphibole
    e. Calcite
    f. Biotite Mica
41. Sample 13: What is this sample?
    a. Magnetite
    b. Graphite
    c. Copper
    d. Sulfur
    e. Hematite
    f. Pyrite
42. Sample 14: What is this sample?
    a. Magnetite
    b. Graphite
    c. Copper
    d. Sulfur
    e. Hematite
    f. Pyrite
43. Sample 15: What is this sample?
    a. Magnetite
    b. Graphite
    c. Copper
    d. Sulfur
    e. Hematite
    f. Pyrite
44. Sample 16: What is this sample?
    a. Magnetite
    b. Graphite
    c. Copper
    d. Sulfur
    e. Hematite
    f. Pyrite
45. Sample 17: What is this sample?
    a. Magnetite
    b. Graphite
    c. Copper
    d. Sulfur
    e. Hematite
    f. Pyrite
46. Sample 18: What is this sample?
    a. Magnetite
    b. Graphite
    c. Copper
    d. Sulfur
    e. Hematite
    f. Pyrite

8 Chapter | IGNEOUS ROCKS

8.1 INTRODUCTION: AN OVERVIEW

All rocks on Earth are classified into one of three categories: igneous, sedimentary, or metamorphic. This rock classification is predicated on the origin of each rock type or, alternatively, on the geological processes that produced the rock. This chapter will concentrate on igneous rocks, the sole type of rock that originates from a molten or liquid state. Consequently, igneous rocks are classified as rock types that originate from the cooling of magma or lava. The classification of igneous rocks is indeed more complex than previously indicated, as there are numerous common varieties, several less prevalent types, and a few that are quite rare, with each igneous rock possessing a distinct name that differentiates it from others within the igneous category. If all originate as molten material (magma or lava) that solidifies to create rock, it is reasonable to conclude that these igneous rocks differ primarily due to: 1) the initial composition of the molten material from which the rock is formed, and 2) the cooling process of the molten material that resulted in the rock's formation. The classification of igneous rocks is determined by two parameters: composition and texture. The composition of igneous rock pertains to its chemical constituents or the minerals contained inside, while the term texture describes observable characteristics, like mineral size, the presence of glass, shattered material, or vesicles (voids) within the volcanic rock.

8.2 IGNEOUS ROCK ORIGIN

8.2.1 Magma Composition

It is paradoxical, yet the formation of igneous rock necessitates the prior existence of molten material, termed magma, which in turn requires a pre-existing rock to be melted, so enabling it to cool and solidify as igneous rock. Which raises more inquiries: which rock underwent melting to generate the magma? Did multiple rock types contribute to the formation of that magma by melting? Did the rocks undergo complete melting, or did only specific minerals inside those rocks experience partial melting? What occurred subsequently to the formation of the melted material? Did an alternative mechanism transpire to alter the composition of that magma prior to its formation as the igneous rock under examination? These are but a selection of inquiries one should contemplate when examining the genesis of igneous rocks.

The majority of rocks, with very few exceptions, consist of minerals that are crystalline solids made up of chemical components. In your chapter on minerals, you discovered that the predominant minerals are classified as silicate minerals, hence it is logical that magmas originate from the melting of rocks that predominantly comprise silicate minerals. All minerals, not exclusively silicates, possess specific conditions, including temperature, at which they can undergo melting. Given that rocks comprise a combination of minerals, it is evident how certain minerals inside a rock may undergo melting while others remain solid. Moreover, temperature conditions are crucial, as only minerals capable of melting at "lower" temperatures (approximately 600°C) may undergo melting, while an increase in temperature (for instance, to 1200°C) is necessary for other minerals to melt as well (noting that the lower temperature minerals continue to melt), thereby contributing their chemical constituents to the generated magma. This raises a significant point: even if identical rock types are undergoing melting, we can produce varying magma compositions solely by melting at various temperatures.

Once created, magma will ascend through the Earth's lithosphere, as it possesses greater buoyancy than the source rock from which it originated. The detachment of magma from its source will lead to altered temperature conditions as it migrates from the heated lithosphere into cooler rock formations, causing the magma to cool as well. Similar to melting, minerals possess specific conditions under which they crystallize from a cooling magma body. The sequence of mineral crystallization is really the reverse of the sequence of crystal melting. Norman L. Bowen experimentally established the sequence of mineral creation from magma in the early 1900s, resulting in the renowned Bowen's reaction series, which is featured in numerous textbooks and laboratory manuals. (Figure 8.1).

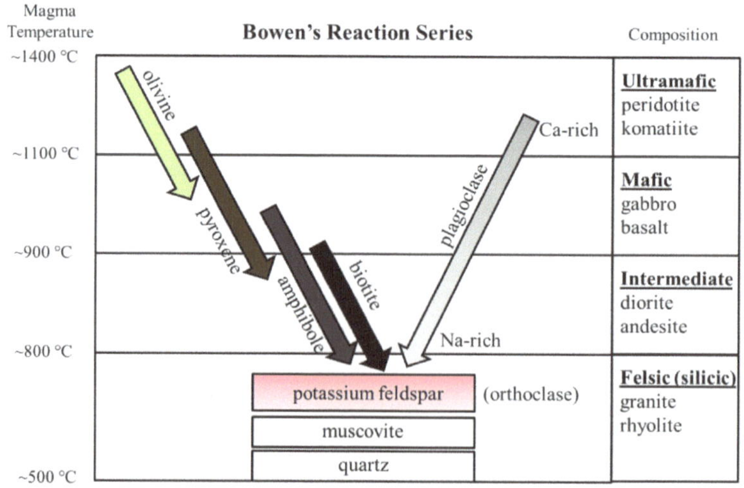

Figure 8.1 | Bowen's Reaction Series, showing the progression of mineral crystallization as magma temperatures drop from ~1400 °C to ~500 °C. Note the

corresponding names for igneous rock composition (underlined and bolded) and some example rock types within each compositional group.

The "reaction series" denotes the sequence of chemical processes involved in mineral formation through the bonding of elements within magma, ordered according to decreasing magma temperatures. A further analysis of Figure 8.1 reveals that olivine is the initial mineral to crystallize from cooling ultramafic magma; the arrow's length denotes the temperature range for olivine formation. When temperatures descend below this threshold, olivine crystals cease to form; rather, other minerals like pyroxene begin to crystallize (there exists a narrow temperature range where both olivine and pyroxene can crystallize simultaneously). Minerals that crystallize from cooling lava are referred to as crystals or phenocrysts; during their formation, these phenocrysts extract chemical components from the magma. Olivine phenocrysts extract magnesium (Mg) and iron (Fe) from the magma and integrate them into their crystalline framework. The propensity of mineral phenocrysts to include specific chemical elements into their structure while omitting others indicates that the composition of the magma must be altering during the formation of phenocrysts.!

There can be more than one mineral type crystallizing within the cooling magma, as the arrows in Figure 8.1 demonstrate. The minerals on the left side of Bowen's reaction series are referred to as a discontinuous series, as these minerals (olivine, pyroxene, amphibole, and biotite) all remove the iron (Fe), magnesium (Mg), and manganese (Mn) from the magma during crystallization, but do so at certain temperature ranges. These iron and magnesium-rich minerals are referred to as **ferromagnesian** minerals (ferro = iron) and are usually green, dark gray, or black in color due to the absorption of visible light by iron and magnesium atoms. On the right side of Bowen's reaction series is a long arrow labelled plagioclase feldspar. Plagioclase crystallizes over a large temperature interval and represents a continuous series of crystallization even though its composition changes from calcium (Ca) rich to sodium (Na) rich. As the magma temperature drops and plagioclase first starts to crystallize (form), it will take in the calcium atoms into the crystal structure, but as magma temperatures continue to drop, plagioclase takes in sodium atoms preferentially. As a result, the higher temperature calcium-rich plagioclase is dark gray in color due to the high calcium content, but the lower temperature sodium-rich plagioclase is white due to the high sodium content. Finally, at the bottom of the graph in Figure 8.1, we see that three more minerals can form as temperatures continue to drop. These minerals (potassium feldspar, muscovite, and quartz) are considered to be the "low temperature minerals", as they are the last to form during cooling, and therefore first to melt as a rock is heated. The prior extraction of iron and magnesium from the magma leads to the emergence of the most recently formed minerals that lack these elements;

these minerals are designated as non-ferromagnesian minerals, characterized by a significantly lighter hue. For instance, potassium-rich feldspar, commonly referred to as orthoclase, may exhibit a pale pink or white hue. References to mineral color are essential, as the hue of any mineral is predominantly determined by its chemical constituents; hence, the color of an igneous rock is contingent upon its mineral content or chemical composition..

8.3 IGNEOUS ROCK COMPOSITION

The Bowen's reaction series diagram frequently includes classes of igneous rocks and corresponding examples, which are fully contingent upon the minerals present inside them. For instance, one might anticipate the presence of copious olivine, together with minor amounts of pyroxene and Ca-rich plagioclase, in ultramafic rocks such as peridotite or komatiite. Conversely, in mafic rocks like gabbro or basalt, pyroxene, plagioclase, and potentially some olivine or amphibole may be found. Quartz, muscovite, potassium feldspar, and possibly little biotite and sodium-rich plagioclase can be anticipated in a felsic (or silicic) rock like granite or rhyolite. Figure 8.1 effectively illustrates that the classification of an igneous rock is somewhat contingent upon the minerals it contains, and given that these minerals exhibit specific colors due to their chemical composition, the rocks consequently possess distinct hues. A rock predominantly made of olivine will exhibit a green hue, attributable to the color of olivine; this type of rock is classified as ultramafic. A rock rich in ferromagnesian minerals would exhibit a dark coloration, as these minerals, except olivine, are typically dark-hued; an igneous rock characterized by a dark color is referred to as a mafic rock, with "ma-" derived from magnesium and "fic" from ferric iron. An igneous rock abundant in non-ferromagnesian minerals will exhibit a light coloration, exemplified by silicic or felsic rocks, with "fel" derived from feldspar and "sic" from silica-rich quartz. Thus, solely based on color, we have commenced the classification of igneous rocks.

Figure 8.2 | Examples of igneous rocks from the mafic (A), intermediate (B), and felsic (C) rock compositions. Photo scale on bottom is in centimeters.

Figure 8.2 illustrates examples of igneous rocks that exemplify mafic and felsic compositions (Figures 8.2A and 8.2C, respectively), together with an intermediate rock type (Figure 8.2B). Felsic rocks may have a little quantity of dark-colored ferromagnesian minerals but are mostly composed of light-colored minerals, whereas mafic rocks possess a greater proportion of dark-colored ferromagnesian minerals, resulting in a darker appearance. An intermediate rock, situated between mafic and felsic classifications, exhibits a mineral composition and coloration that reflects this status; it contains fewer ferromagnesian minerals than mafic rocks, yet more than felsic rocks.

Classifying rocks into igneous compositions (ultramafic, mafic, intermediate, and felsic) is contingent upon the minerals present in each rock. Identifying minerals in rocks, as depicted in Figure 8.2A, can be challenging due to the predominance of dark-colored minerals, which complicates their differentiation. A straightforward approach to ascertain the composition of igneous rock involves calculating the percentage of dark-hued minerals, without identifying the specific minerals present; this classification method utilizes a mafic color index (MCI), with "mafic" denoting any dark gray, black, or green mineral (Figure 8.3).

Figure 8.3 | Examples of how igneous rocks can be classified using the Mafic Color Index (MCI), which is a visual classification based on the amount of ferromagnesian minerals in the rock: (A) The small amount of tiny black phenocrysts (biotite) gives this rock a 0-15% MCI value; (B) numerous dark phenocrysts (amphibole) gives this rock a 16-45% MCI value; (C) this rock lacks visible phenocrysts, but the black color in this rock results in a 46-85% MCI value; (D) this rock is entirely green in color due to the overwhelming amount of olivine. Any rock with this much olivine is always classified as having over 85% MCI values.

Igneous rocks containing 0-15% dark-colored minerals (or 0-15% MCI) are classified as felsic rocks (Figure 8.3A), those with 46-85% MCI are categorized as mafic rocks (Figure 8.3C), and rocks with over 85% MCI are designated as ultramafic (Figure 8.3D). Rocks with an intermediate composition or a 16-45% MCI are classified as intermediate igneous rocks (Figure 8.3B). Determining the percentage of dark-colored minerals is feasible only when the minerals are sufficiently large for visibility; in such instances, one can identify a mafic rock by its dark hue and a felsic rock by its light hue. An intermediate rock is somewhat lighter than a mafic rock but darker than a felsic rock. Ultimately, an ultramafic rock is characteristically green, owing to its substantial content of green-hued olivine. Rocks containing minerals that are imperceptible to the naked eye are illustrated in Figure 8.4; it is important to observe that distinctions may still be made between mafic (Figure 8.4A), intermediate (Figure 8.4B), and felsic (Figure 8.4C) based on the overall coloration of the rock. The intermediate igneous rock depicted in Figure 8.4B contains several discernible phenocrysts; this unusual texture will be addressed later in this chapter.

Figure 8.4 | Examples of igneous rocks from the mafic (A), intermediate (B), and felsic (C) rock compositions. Notice the difference in appearance between these rocks and those in Figure 8.2 Author: Karen Tefend

8.3.1 LAB EXERCISE

Part A Igneous Rock Composition

Before attempting to answer the following questions, remove the eight rock samples from the Igneous Rocks bag in your HOL rock kit (samples I1 – I8) and place them on a clean sheet of white paper. Your samples should look identical to the samples in Figure 8.5.

Separate the 8 samples into three groups based on their color (dark, light or intermediate). As a first attempt at classifying rock compositions based on color, most students end up with three light colored rocks (felsic), two intermediate rocks, and three dark colored rocks (mafic). There actually are only two mafic, two intermediate, two felsic, and two other rocks that we have yet to discuss, but for now leave your samples as they are. Take a close look at the dark colored rocks in your mafic pile; how are they different? When you hold them closer to a light source, notice how they reflect light; one is dull looking, one has small shiny surfaces, and one is extremely shiny and smooth. The one that is very smooth and shiny is not a mafic rock, even though it is dark in color. This particular rock is an example of that rare rock type that lacks minerals (crystals); instead, it is almost entirely made of glass, and is most probably felsic in composition. Go ahead and move this rock to its own space on your piece of paper. Now take a close look at your three light-colored felsic rocks. One will have several minerals visible that you will see as different colors. The other two rocks will be fairly uniform in color; one is composed of very tiny minerals that require a microscope to view them, and the other rock seems to be fragile and lightweight. This fragile appearing rock is another example of rock that is again almost entirely glass with a very few, tiny phenocrysts; set this rock aside with your other glassy rock. Now you should have four piles, with two rocks in each pile, and you can now proceed with the following questions. You may want to refer to Figures 8.1 through 8.4 to help with identification.

Figure 8.5 | These samples are in your rock kit, in a bag labeled "IGNEOUS ROCKS". Each sample is identified by the letter "I" for igneous, and a number 1 through 8. A hand lens is also provided in your rock kit; use the hand lens for magnified viewing.

1. Sample I1 has what mafic color index (MCI)?

   a. 0-15% MCI b. 16-45% MCI c. 46-85% MCI d. >85% MCI

2. Sample I1 has pink colored minerals present, which aremost likely___.

   a. quartz b. biotite c. orthoclase d. Ca-rich plagioclase

3. The small, dark colored minerals in Sample I1 are most likely_.

   a. quartz b. biotite c. orthoclase d. Na-rich plagioclase

4. Sample I1 belongs to which igneous rock classification?

   a. ultramafic b. mafic c. intermediate d. felsic

5. Sample I2 has what mafic color index (MCI)?

   a. 0-15% MCI b. 16-45% MCI c. 46-85% MCI d. >85% MCI

6. Sample I2 belongs to which igneous rock classification?

   ultramafic b. mafic c. intermediate d. felsic

7. Sample I3 has what mafic color index (MCI)?

   a. 0-15% MCI b. 16-45% MCI c. 46-85% MCI d. >85% MCI

8. Sample I3 belongs to which igneous rock classification?

   a. ultramafic b. mafic c. intermediate d. felsic

9. Sample I4 has what mafic color index (MCI)?

   a. 0-15% MCI b. 16-45% MCI c. 46-85% MCI d. >85% MCI

10. Sample I4 belongs to which igneous rock classification?

    a. ultramafic b. mafic c. intermediate d. felsic

11. Sample I5 has what mafic color index (MCI)?

    a. 0-15% MCI b. 16-45% MCI c. 46-85% MCI d. >85% MCI

12. The white minerals in Sample I5 are most likely .

    a. quartz b. biotite c. orthoclase d. plagioclase

13. Sample I7 has what mafic color index (MCI)?

    a. 0-15% MCI b. 16-45% MCI c. 46-85% MCI d. >85% MCI

14. According to Figure 8.1, Sample I4 will never contain which mineral?

    a. olivine b. pyroxene c. amphibole d. quartz

15. According to Figure 8.1, Sample I7 must contain which mineral?

    a. olivine b. pyroxene c. amphibole d. quartz

16. According to Bowen's Reaction Series, Sample I5 (if heated) should start melting at what temperature?

    a. ~1100 °C b. ~900 °C c. ~800 °C d. ~500 °C

17. According to Figure 8.1, when Sample I1 is heated, what is the first mineral to melt?

    a. olivine b. Ca-rich plagioclase

    c. potassium feldspar d. quartz

8.4 IGNEOUS ROCK TEXTURE

The categorization of igneous rocks relies on both composition and texture. Texture pertains to the characteristics observed in rock, including mineral sizes and the presence of glass, shattered material, or vesicles (voids) in igneous rock. This section will address mineral crystal dimensions and vesicles.

The size of the crystals or phenocrysts, which develop during the cooling of magma, is intrinsically linked to the cooling process. Each mineral's chemical makeup originates from the magma, and each mineral possesses a certain temperature range within which it can crystallize. The chemical elements that constitute the mineral must migrate from the liquid magma to connect with other elements in a specific manner, so forming the distinctive crystal structure of that mineral. What are the potential outcomes if the temperature of the magma decreases rapidly, or if it decreases gradually? The duration permitted for the migration of chemical components to crystallize is influenced. When magma cools gradually, there is ample opportunity for the movement of essential chemical elements to crystallize into a specific mineral; that mineral can attain considerable size, sufficient for visibility without a microscope. This igneous rock, characterized by its discernible minerals, is described as possessing a phaneritic texture (phan = big). The rock specimens depicted in Figure 8.2 are exclusively phaneritic rocks. Figure 8.2A depicts a phaneritic mafic rock identified as gabbro, Figures 8.2B and 8.3B illustrate a phaneritic intermediate rock referred to as diorite, while the rock in Figure 8.2C is a phaneritic felsic rock recognized as granite. Upon consulting Figure 8.1 (Bowen's reaction series), you will observe that these rock names are enumerated on the right side of the diagram.

Magma that cools rapidly will yield the contrary outcome; there is insufficient time for the movement of chemical elements to create minerals, resulting in the formation of smaller crystals. Consequently, several tiny crystals of a certain mineral will develop within the magma. Igneous rocks with crystals that are imperceptible to the naked eye, save under a microscope, are referred to as aphanitic igneous rocks. Figures 8.3C, 8.4A, and 8.4C represent aphanitic rocks; specifically, Figure 8.3C (and 8.4A) is characterized by its dark coloration, identifying it as a mafic aphanitic rock known as basalt. The felsic rock depicted in Figure 8.4C is referred to as rhyolite. Basalt and gabbro are both mafic rocks with identical composition; however, basalt signifies magma that cooled rapidly, while gabbro denotes mafic magma that cooled slowly. Similarly, the other rock compositions exhibit same phenomenon: the felsic rocks rhyolite and granite possess identical compositions, although one cooled rapidly (rhyolite) and the other cooled gradually (granite). The intermediate rocks diorite (Figure 8.2B) and andesite (Figure 8.4B) signify magmas that cooled at a modest or moderately rapid rate, respectively. Occasionally, visible crystals are present in an otherwise aphanitic rock, exemplified by the andesite in Figure 8.4B. The rock's texture is termed porphyritic, or more precisely porphyritic-aphanitic, as it is a porphyritic andesite, and all andesites are aphanitic. The presence of two distinct crystal sizes in an igneous rock signifies an increase in the magma's cooling rate; larger crystals develop during slow cooling, whereas rapid cooling results in the formation of smaller crystals. A phaneritic rock may also be designated as a porphyritic-phaneritic rock if it contains notably large crystals, comparable in size to a thumb, alongside other visible crystals. Figure 8.6 illustrates two porphyritic rocks: a porphyritic-aphanitic basalt and a porphyritic-phaneritic granite.

Figure 8.6 | (A) An example of a porphyritic-aphanitic mafic rock with needle-shaped amphibole phenocrysts (arrow points to one phenocryst that is 1cm in length); No other minerals in (A) are large enough to see. (B) An example of a porphyritic-phaneritic felsic rock with large feldspars (outlined phenocryst is 3 cm length). Surrounding these large feldspars are smaller (yet still visible) dark and light colored minerals.

Occasionally, magma cools so rapidly that there is insufficient time for mineral formation, as the chemical constituents within the magma cannot migrate into a crystalline structure. When this occurs, the magma transforms into a dense glass known as obsidian (Figure 8.7A). Glass is defined as a disordered arrangement of chemical elements, so it is not classified as a mineral; igneous rocks predominantly constituted of glass exhibit a glassy texture. Identifying a glassy rock like obsidian is straightforward if you remember the characteristics of glass; any thick glass pane or a shattered glass bottle will exhibit a smooth, curved pattern on the fractured edge known as conchoidal fracture (as discussed in your mineral chapter). Although obsidian is a naturally occurring substance rather than a synthetic one, it nonetheless fractures in a conchoidal pattern. Upon close examination of the obsidian in Figure 8.7A, one may observe the curved (conchoidal) surfaces by recognizing the lustrous pattern on the rock. Obsidian exhibits a consistently dark hue irrespective of its composition due to its opaque glass structure, which obstructs light transmission; nevertheless, if the margins of the obsidian specimen are sufficiently thin, translucence may be observed.

Figure 8.7 | Igneous rocks with glassy texture: obsidian (A) and pumice (B).

In Figure 8.7B there is another igneous rock that is also composed primarily of glass due to a very fast rate of magma cooling. This rock is called pumice, and is commonly referred to as the rock that floats on water due to its low density. The glass in this rock is stretched out into very fine fibers of glass which formed during the eruptive phase of a volcano. Because these fibers are so thin, they are easy to break (unlike the dense obsidian) and any conchoidal fractures on these fibers are too small to see without the aid of a microscope. Pumice can have any composition (felsic to mafic), but unlike obsidian the color of the pumice can be used to determine the magma composition, as felsic pumice is always light in color and mafic pumice will be dark in color. Mafic pumice with a dark grey, red or black color is also known as scoria. 8.6 IGNEOUS ROCK FORMATION—PLUTONIC VS. VOLCANIC

The different crystal sizes and presence or absence of glass in an igneous rock is primarily controlled by the rate of magma cooling. Magmas that cool below the surface of the earth tend to cool slowly, as the surrounding rock acts as an insulator, not unlike a coffee thermos. Most of us are aware that even the most expensive coffee thermos does not prevent your coffee from cooling; it just slows the rate of cooling. Magma that stays below the surface of the earth can take tens of thousands of years to completely crystallize depending on the size of the magma body. Once the magma has completely crystallized, the entire igneous body is called a pluton. Sometimes portions of these plutons are exposed at the earth's surface where direct observation of the rock is possible, and upon inspection of this "plutonic" rock, you would see that it is composed of minerals that are large enough to see without the aid of a microscope. Therefore, any igneous rock sample that is considered to have a phaneritic texture (or porphyritic-phaneritic), is also referred to as a **plutonic** rock. A plutonic rock is also called an **intrusive** rock as it is derived from magma that intruded the rock layers but never reached the earth's surface.

If magma does manage to reach the earth's surface, it is no longer insulated by the rocks around it and will therefore cool rapidly. Magma that reaches the earth's surface through a fissure or central vent (volcano) will lose some of its dissolved gas and becomes lava, and any rock that forms from lava will have an aphanitic texture due to fast cooling (or have a glassy texture due to very fast cooling). Flowing lava may continue to release gas while

cooling; this is typical of mafic lava flows. If the lava hardens while these gases are bubbling out of the lava, a small hole or vesicle may form on the rock, and even though the rock is aphanitic (due to fast cooling), the name of the rock can be given the term "**vesicular**" to indicate the presence of these vesicles. For example, a basalt with vesicles is called vesicular basalt (Figure 8.8). Aphanitic rocks and rocks with glassy texture are also

known as **volcanic** igneous rocks, or **extrusive** igneous rocks, as the magma was "extruded" onto the surface of the earth. Porphyritic-aphanitic rocks are also considered to be volcanic or extrusive rocks; these rocks may have begun crystallizing under the earth's surface as a slowly cooling magma (as evidence by the small amount of visible phenocrysts), but this magma was extruded onto the surface as lava before finally crystallizing completely on the surface of the earth to form an extrusive igneous rock with a porphyritic texture, or more accurately, a porphyritic-aphanitic texture.

Figure 8.8 | An aphanitic mafic rock (basalt), with gas

escape structures called vesicles. Arrow points to one vesicle that is ~1cm in diameter. This is an example of another texture type, called vesicular texture, and the name of this rock is a vesicular basalt.

Figure 8.9 presents a review of the terminology employed to categorize igneous rocks, facilitating the identification of the igneous rock samples included in your rock kit. Consult the aforementioned figures for additional assistance.

Composition

Texture	Felsic	Intermediate	Mafic
	0-15% MCI	16-45% MCI	46-85% MCI
Phaneritic visible minerals, slow cooling (plutonic)	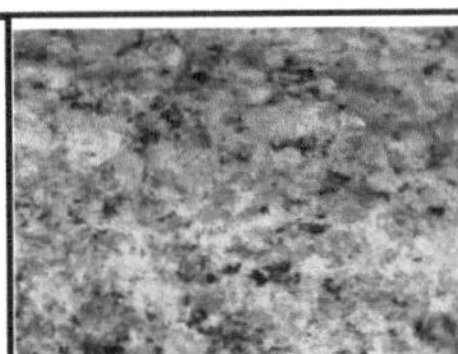 granite	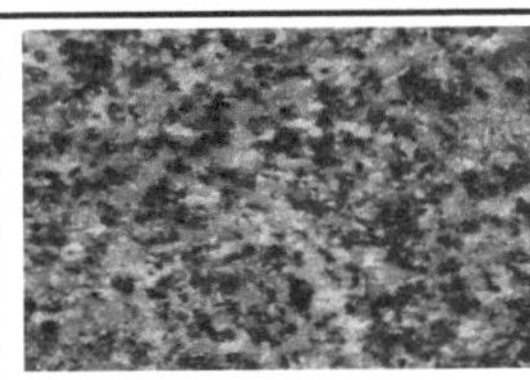 diorite	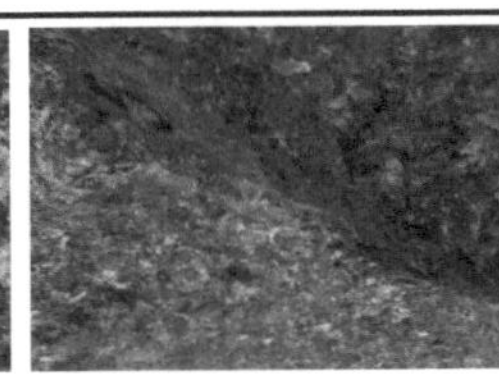 gabbro
Aphanitic No visible crystals, fast cooling (volcanic)	 rhyolite	 andesite*	 basalt
Porphyritic	Use this term if there are visible crystals in basalt or andesite, or if there are very large crystals in granite		
Glassy	Dense glass that is dark in color (obsidian), or Glass froth that is light or intermediate in color (pumice)		

*this andesite is porphyritic-aphanitic.

Figure 8.9 | Chart showing some common igneous rock textures and compositions. MCI = mafic color index, or the percentage of dark colored ferromagnesian minerals present. Recall that any composition can be phaneritic, aphanitic, porphyritic or glassy. Vesicular texture is not as common and is only seen in some aphanitic rocks.

8.4.1 LAB EXERCISE

Part B Igneous Rock Texture

For this exercise, you will determine the texture and rate of cooling for the igneous rock samples in your HOL kit. Refer to Figures 8.6 through 8.9 for guidance.

18. Sample I1 has what texture?
    a. phaneritic
    b. aphanitic
    c. porphyritic-aphanitic
    d. glassy
19. How did Sample I1 form?
    a. by magma that cooled slowly
    b. by fast cooling lava
    c. by a very fast cooling of a lava
    d. by a very fast cooling magma during a volcanic eruption
    e. by magma that was cooling slowly, then as a lava that cooled quickly
20. Sample I1 is called__.
    a. obsidian
    b. pumice
    c. granite
    d. rhyolite
    e. diorite
    f. andesite
    g. gabbro
    h. basalt
21. Sample I1 can also be called_.
    a. volcanic
    b. plutonic
22. Sample I2 has what texture?
    a. phaneritic
    b. aphanitic
    c. porphyritic-aphanitic
    d. glassy
23. How did Sample I2 form?
    a. by magma that cooled slowly
    b. by fast cooling lava
    c. by a very fast cooling of a lava
    d. by a very fast cooling magma during a volcanic eruption
    e. by magma that was cooling slowly, then as a lava that cooled quickly
24. Sample I2 is called__.
    a. obsidian
    b. pumice
    c. granite
    d. rhyolite
    e. diorite
    f. andesite
    g. gabbro
    h. basalt
25. Sample I2 can also be called_.
    a. volcanic
    b. plutonic

26. Sample I3 has what texture?
    a. phaneritic
    b. aphanitic
    c. porphyritic-aphanitic
    d. glassy
27. How did Sample I3 form?
    a. by magma that cooled slowly
    b. by fast cooling lava
    c. by a very fast cooling of a lava
    d. by a very fast cooling magma during a volcanic eruption
    e. by magma that was cooling slowly, then as a lava that cooled quickly
28. Sample I3 is called__.
    a. obsidian
    b. pumice
    c. granite
    d. rhyolite
    e. diorite
    f. andesite
    g. gabbro
    h. basalt
29. Sample I3 can also be called_.
    a. volcanic
    b. plutonic
30. Sample I4 has what texture?
    a. phaneritic
    b. aphanitic
    c. porphyritic-aphanitic
    d. glassy
31. How did Sample I4 form?
    a. by magma that cooled slowly
    b. by fast cooling lava
    c. by a very fast cooling of a lava
    d. by a very fast cooling magma during a volcanic eruption
    e. by magma that was cooling slowly, then as a lava that cooled quickly
32. Sample I4 is called__.
    a. obsidian
    b. pumice
    c. granite
    d. rhyolite
    e. diorite
    f. andesite
    g. gabbro
    h. basalt
33. Sample I4 can also be called_.
    A. volcanic
    b. plutonic

34. Sample I5 has what texture?
    - a. phaneritic
    - b. aphanitic
    - c. porphyritic-aphanitic
    - d. glassy
35. How did Sample I5 form?
    - a. by magma that cooled slowly
    - b. by fast cooling lava
    - c. by a very fast cooling of a lava
    - d. by a very fast cooling magma during a volcanic eruption
    - e. by magma that was cooling slowly, then as a lava that cooled quickly
36. Sample I5 is called .
    - a. obsidian
    - b. pumice
    - c. granite
    - d. rhyolite
    - e. diorite
    - f. andesite
    - g. gabbro
    - h. basalt
37. Sample I5 can also be called .
    - a. volcanic
    - b. plutonic
38. Sample I6 has what texture?
    - a. phaneritic
    - b. aphanitic
    - c. porphyritic-aphanitic
    - d. glassy
39. Sample I6 has what composition?
    - a. felsic
    - b. intermediate
    - c. mafic
    - d. any of the above
40. How did Sample I6 form?
    - a. by magma that cooled slowly
    - b. by fast cooling lava
    - c. by a very fast cooling of a lava
    - d. by a very fast cooling magma during a volcanic eruption
    - e. by magma that was cooling slowly, then as a lava that cooled quickly
41. Sample I6 is called_.
    - a. obsidian
    - b. pumice
    - c. granite
    - d. rhyolite
    - e. diorite
    - f. andesite
    - g. gabbro
    - h. basalt

42. Sample I7 has what texture?
    a. phaneritic
    b. aphanitic
    c. porphyritic-aphanitic
    d. glassy

43. How did Sample I7 form?
    a. by magma that cooled slowly
    b. by fast cooling lava
    c. by a very fast cooling of a lava
    d. by a very fast cooling magma during a volcanic eruption
    e. by magma that was cooling slowly, then as a lava that cooled quickly

44. Sample I7 is called__.
    a. obsidian
    b. pumice
    c. granite
    d. rhyolite
    e. diorite
    f. andesite
    g. gabbro
    h. basalt

45. Sample I7 can also be called_.
    a. volcanic
    b. plutonic

46. Sample I8 has what texture?
    a. phaneritic
    b. aphanitic
    c. porphyritic-aphanitic
    d. glassy

47. Sample I8 has what composition?
    a. felsic
    b. intermediate
    c. mafic
    d. any of the above

48. How did Sample I8 form?
    a. by magma that cooled slowly
    b. by fast cooling lava
    c. by a very fast cooling of a lava
    d. by a very fast cooling magma during a volcanic eruption
    e. by magma that was cooling slowly, then as a lava that cooled quickly

49. Sample I8 is called__.
    a. obsidian
    b. pumice
    c. granite
    d. rhyolite
    e. diorite
    f. andesite
    g. gabbro
    h. basalt

Part C Google Earth

Igneous intrusions such as plutons are large magma bodies that have crystallized into solid rock. Regional uplift and erosion of the earth's surface can expose portions of these plutons, and large exposures that are over 100 km2 are called batholiths, and a small portion (less than 100 km2) of a pluton exposed at the earth's surface is called a stock. The following exercises require that you use Google Earth.

50. Type the following coordinates into the search bar in Google Earth: 33 48 14.24 N 84 08 44.31 W. This is Stone Mountain, GA. Zoom in to an eye elevation of ~2400 ft for a closer look. Based on the color, and assuming no color change has occurred due to weathering at this location, what is the composition of rock?

    a. ultramafic b. mafic

    c. intermediate d. felsic

51. Zoom out to an eye elevation of ~1500 ft so that you can see the entire mountain. Click on the ruler icon to open up the ruler function. Select the Line tab, and change the map length to kilometers. What is the longest length of exposed portion of Stone Mountain?

    a. 3000 km b. 260 km

    c. 6 km d. 2.6 km

52. Based on your measurement of the longest dimension, Stone Mountain can be classified as:

    a batholith b. a stock

53. Type the following coordinates into the search bar in Google Earth: 64 58 45.54 N 16 43 07.04 W. Zoom in to an eye elevation of ~12979 ft for a closer look. Based on the color, what is the composition of this rock?

    a. ultramafic b. mafic

    c. intermediate d. felsic

54. Zoom out to an eye elevation of 41,566 ft. Based on the proximity of the volcanic crater (now containing a lake), the igneous rocks identified in question 53 are:

    a. extrusive b. intrusive

8.5 STUDENT RESPONSES

1. Sample I1 has what mafic color index (MCI)?

    a. 0-15% MCI b. 16-45% MCI

    c. 46-85% MCI d. >85% MCI

2. Sample I1 has pink colored minerals present, which aremost likely____.

    a. quartz b. biotite

    c. orthoclase d. Ca-rich plagioclase

3. The small, dark colored minerals in Sample I1 are most likely_.
   a. quartz b. biotite
   c. orthoclase d. Na-rich plagioclase

4. Sample I1 belongs to which igneous rock classification?
   a. ultramafic b. mafic
   c. intermediate d. felsic

5. Sample I2 has what mafic color index (MCI)?
   a. 0-15% MCI b. 16-45% MCI
   c. 46-85% MCI d. >85% MCI

6. Sample I2 belongs to which igneous rock classification?
   a. ultramafic b. mafic
   c. intermediate d. felsic

7. Sample I3 has what mafic color index (MCI)?
   a. 0-15% MCI b. 16-45% MCI
   c. 46-85% MCI d. >85% MCI

8. Sample I3 belongs to which igneous rock classification?
   a. ultramafic b. mafic
   c. intermediate d. felsic

9. Sample I4 has what mafic color index (MCI)?
   a. 0-15% MCI b. 16-45% MCI
   c. 46-85% MCI d. >85% MCI

10. Sample I4 belongs to which igneous rock classification?
    a. ultramafic b. mafic
    c. intermediate d. felsic

11. Sample I5 has what mafic color index (MCI)?
    a. 0-15% MCI b. 16-45% MCI
    c. 46-85% MCI d. >85% MCI

12. The white minerals in Sample I5 are most likely .
    a. quartz b. biotite
    c. orthoclase d. plagioclase

13. Sample I7 has what mafic color index (MCI)?
    a. 0-15% MCI b. 16-45% MCI
    c. 46-85% MCI d. >85% MCI

14. According to Figure 8.1, Sample I4 will never contain which mineral?
    a. olivine
    b. pyroxene
    c. amphibole
    d. quartz
15. According to Figure 8.1, Sample I7 must contain which mineral?
    a. olivine
    b. pyroxene
    c. amphibole
    d. quartz
16. According to Bowen's Reaction Series, Sample I5 (if heated) should start melting at what temperature?
    a. ~1100 °C
    b. ~900 °C
    c. ~800 °C
    d. ~500 °C
17. According to Figure 8.1, when Sample I1 is heated, what is the first mineral to melt?
    a. olivine
    b. Ca-rich plagioclase
    c. potassium feldspar
    d. quartz
18. Sample I1 has what texture?
    a. phaneritic
    b. aphanitic
    c. porphyritic-aphanitic
    d. glassy
19. How did Sample I1 form?
    a. by magma that cooled slowly
    b. by fast cooling lava
    c. by a very fast cooling of a lava
    d. by a very fast cooling magma during a volcanic eruption
    e. by magma that was cooling slowly, then as a lava that cooled quickly
20. Sample I1 is called_.
    a. obsidian
    b. pumice
    c. granite
    d. rhyolite
    e. diorite
    f. andesite
    g. gabbro
    h. basalt
21. Sample I1 can also be called_.
    a. volcanic
    b. plutonic
22. Sample I2 has what texture?
    a. phaneritic
    b. aphanitic\
    c. porphyritic-aphanitic
    d. glassy

23. How did Sample I2 form?
    a. by magma that cooled slowly
    b. by fast cooling lava
    c. by a very fast cooling of a lava
    d. by a very fast cooling magma during a volcanic eruption
    e. by magma that was cooling slowly, then as a lava that cooled quickly

24. Sample I2 is called_.
    a. obsidian b. pumice
    c. granite d. rhyolite
    e. diorite f. andesite
    g. gabbro h. basalt

25. Sample I2 can also be called_.
    a. volcanic b. plutonic

26. Sample I3 has what texture?
    a. phaneritic b. aphanitic
    c. porphyritic-aphanitic d. glassy

27. How did Sample I3 form?
    a. by magma that cooled slowly
    b. by fast cooling lava
    c. by a very fast cooling of a lava
    d. by a very fast cooling magma during a volcanic eruption
    e. by magma that was cooling slowly, then as a lava that cooled quickly

28. Sample I3 is called_.
    a. obsidian b. pumice
    c. granite d. rhyolite
    e. diorite f. andesite
    g. gabbro h. basalt

29. Sample I3 can also be called_.
    a. volcanic b. plutonic

30. Sample I4 has what texture?
    a. phaneritic b. aphanitic
    c. porphyritic-aphanitic d. glassy

31. How did Sample I4 form?
    a. by magma that cooled slowly
    b. by fast cooling lava
    c. by a very fast cooling of a lava
    d. by a very fast cooling magma during a volcanic eruption
    e. by magma that was cooling slowly, then as a lava that cooled quickly
32. Sample I4 is called__.
    a. obsidian b. pumice
    c. granite d. rhyolite
    e. diorite f. andesite
    g. gabbro h. basalt
33. Sample I4 can also be called_.
    a. volcanic b. plutonic
34. Sample I5 has what texture?
    a. phaneritic b. aphanitic
    c. porphyritic-aphanitic d. glassy
35. How did Sample I5 form?
    a. by magma that cooled slowly
    b. by fast cooling lava
    c. by a very fast cooling of a lava
    d. by a very fast cooling magma during a volcanic eruption
    e. by magma that was cooling slowly, then as a lava that cooled quickly
36. Sample I5 is called .
    a. obsidian b. pumice
    c. granite d. rhyolite
    e. diorite f. andesite
    g. gabbro h. basalt
37. Sample I5 can also be called .
    a. volcanic b. plutonic
38. Sample I6 has what texture?
    a. phaneritic b. aphanitic
    c. porphyritic-aphanitic d. glassy

39. Sample I6 has what composition?

a. felsic b. intermediate

c. mafic d. any of the above

40. How did Sample I6 form?

a. by magma that cooled slowly

b. by fast cooling lava

c. by a very fast cooling of a lava

d. by a very fast cooling magma during a volcanic eruption

e. by magma that was cooling slowly, then as a lava that cooled quickly

41. Sample I6 is called_.

a. obsidian b. pumice

c. granite d. rhyolite

e. diorite f. andesite

g. gabbro h. basalt

42. Sample I7 has what texture?

a. phaneritic b. aphanitic

c. porphyritic-aphanitic d. glassy

43. How did Sample I7 form?

a. by magma that cooled slowly

b. by fast cooling lava

c. by a very fast cooling of a lava

d. by a very fast cooling magma during a volcanic eruption

e. by magma that was cooling slowly, then as a lava that cooled quickly

44. Sample I7 is called_.

a. obsidian b. pumice

c. granite d. rhyolite

e. diorite f. andesite

g. gabbro h. basalt

45. Sample I7 can also be called_.

a. volcanic b. plutonic

46. Sample I8 has what texture?

a. phaneritic b. aphanitic

c. porphyritic-aphanitic d. glassy

47. Sample I8 has what composition?
    a. felsic
    b. intermediate
    c. mafic
    d. any of the above
48. How did Sample I8 form?
    a. by magma that cooled slowly
    b. by fast cooling lava
    c. by a very fast cooling of a lava
    d. by a very fast cooling magma during a volcanic eruption
    e. by magma that was cooling slowly, then as a lava that cooled quickly
49. Sample I8 is called__.
    a. obsidian
    b. pumice
    c. granite
    d. rhyolite
    e. diorite
    f. andesit
    g. gabbro
    h. basalt
50. Type the following coordinates into the search bar in Google Earth: 33 48 14.24 N 84 08 44.31 W. This is Stone Mountain, GA. Zoom in to an eye elevation of ~2400 ft for a closer look. Based on the color, and assuming no color change has occurred due to weathering at this location, what is the composition of rock?
    a. ultramafic
    b. mafic
    c. intermediate
    d. felsic
51. Zoom out to an eye elevation of ~1500 ft so that you can see the entire mountain. Click on the ruler icon to open up the ruler function. Select the Line tab, and change the map length to kilometers. What is the longest length of exposed portion of Stone Mountain?
    a. 3000 km
    b. 260 km
    c. 6 km
    d. 2.6 km
52. Based on your measurement of the longest dimension, Stone Mountain can be classified as:
    a batholith
    b. a stock
53. Type the following coordinates into the search bar in Google Earth: 64 58 45.54 N 16 43 07.04 W. Zoom in to an eye elevation of ~12979 ft for a closer look. Based on the color, what is the composition of this rock?
    a. ultramafic
    b. mafic
    c. intermediate
    d. felsic
54. Zoom out to an eye elevation of 41,566 ft. Based on the proximity of the volcanic crater (now containing a lake), the igneous rocks identified in question 53 are:
    a. extrusive
    b. intrusive

9 Chapter | VOLCANOES

9.1 INTRODUCTION: AN OVERVIEW

What are your thoughts about residing on an active volcano? Many individuals reside on or adjacent to active volcanoes, while several more inhabit areas near volcanoes deemed "dormant," which have experienced a volcanically quiescent phase for the preceding 10,000 years yet possess the capacity to erupt in the future (Figure 9.1). Are they insane (Figure 9.2)? While some volcanoes exhibit explosive eruptions, not all do; for instance, the volcanoes that constitute the Hawaiian Islands are characterized by effusive eruptions, with lava cascading down their flanks (Figure 9.3). Hawaiian residents are acquainted with this eruption style and recognize that their specific volcano will not explode explosively. Volcanic eruptions are more probable in the volcanoes of the Islands of Sumatra and Java (Indonesia); these volcanoes pose significant hazards, although Indonesia is among the most densely populated nations globally. It is crucial to understand the reasons some volcanoes are hazardous while others are not. Will we have sufficient warning before the currently dormant "dangerous" kinds become active? Should we be concerned solely with the eruption, or are there additional threats we need to consider? This chapter will examine these topics, along with additional inquiries regarding the formation of volcanoes in specific regions, particularly in relation to plate tectonics, and the correlation between magma type, volcano type, and eruptive style.

Figure 9.1 | An idyllic setting with rolling fields, crashing waves, and a beautiful snow-capped volcano is pictured.

Figure 9.2 | A not-so-idyllic photo of a volcano during an explosive eruption is pictured.

Figure 9.3 | Nighttime photo of an erupting volcano; note that this particular volcanic eruption involves the flow of lava down the sides of the volcano, as opposed to the explosive eruption in the previous image, Figure 9.2.

9.2 MAGMA GENERATION

This section will examine the concepts of partial melting from the preceding chapter on igneous rocks and the requirements for mantle rock melting discussed in the chapter on plate tectonics.

The majority of magma is produced at the base of the Earth's crust; Figure 9.4 presents a pressure-temperature curve akin to the one encountered in the plate tectonics chapter. To the left of the solid black line, known as the solidus, is a zone where the temperature is insufficient for rock to undergo melting. To the right of the solidus line lies the area where rock undergoes melting.

Observe that the solidus line is not a vertical line descending directly, but rather inclined at an inclination less than vertical, indicating that as pressure increases, the temperature must also rise for a rock to undergo melting. Examine the conditions at the base of the crust, specifically at position "X". The rock at point X lacks sufficient heat to melt, or alternatively, it is subjected to excessive pressure to undergo melting. To induce the melting of this rock, an elevation in temperature (arrow "a"), a reduction in pressure (arrow "b"), or a simultaneous occurrence of both elevated temperature and diminished pressure (arrow "c") is required.

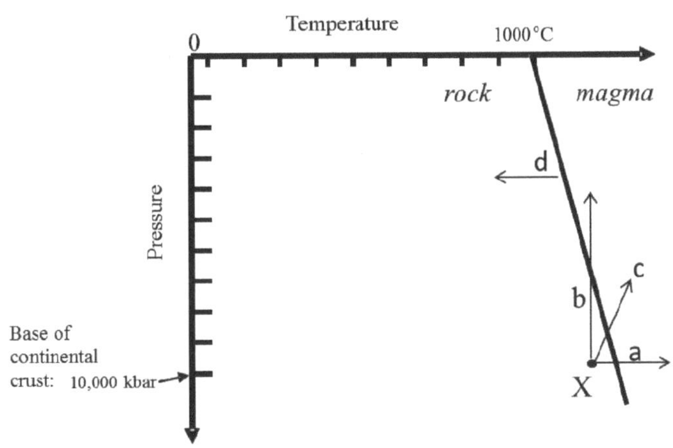

Figure 9.4 | Pressure and Temperature graph of crustal rocks.

Regardless of the path taken, we can make this rock X cross the solid line and become magma. The only other way we can make rock X cross the solid line and become magma is to move this line (arrow "d" on Figure 9.4); in other words, change the melting temperature of the rock. This can be done by adding water, which lowers the melting temperature of rock, and now we can make rock X melt without actually having to change the temperature and pressure conditions.

Now let us think of plate tectonics and the types of boundaries that have magma associated with them (Figure 9.5). Tectonic plates that are diverging (or pulling apart), causes the underlying region of the mantle to experience reduced pressure conditions (just like what the cheerleader at the bottom of a pyramid experiences when everyone else jumps off his back). If the mantle is already fairly warm, the decreased pressure may just be enough for magma to be produced (arrow "b" in Figure 9.5). Where tectonic plates are converging (coming together), one of the plates may subduct below the other plate; recall that subduction will only occur if the tectonic plate has an oceanic crust type. This subducting oceanic crust-topped plate will contain minerals that are hydrated (water in their crystal structure), and as the plate subducts, the hydrated minerals will become unstable and water will be released.

This water will lower the melting temperature of the mantle region directly above the subducting plate, and as a result magma is produced (arrow "d" in Figure

9.5). The last way to melt rock is to just increase the temperature of the rock; this particular melting mechanism does not have to be associated with any particular plate boundary. Instead there must be a region known as a hot spot, caused by mantle plumes (arrow "a" in Figure 9.5). Mantle plumes are thought to be generated at the core-mantle boundary, and are regions of increased temperature that can cause melting of the lithospheric region. With the lithosphere broken into several tectonic plates that have been migrating over these plumes throughout geologic time, the resulting hotspot-generated volcanoes can be found anywhere in the world.

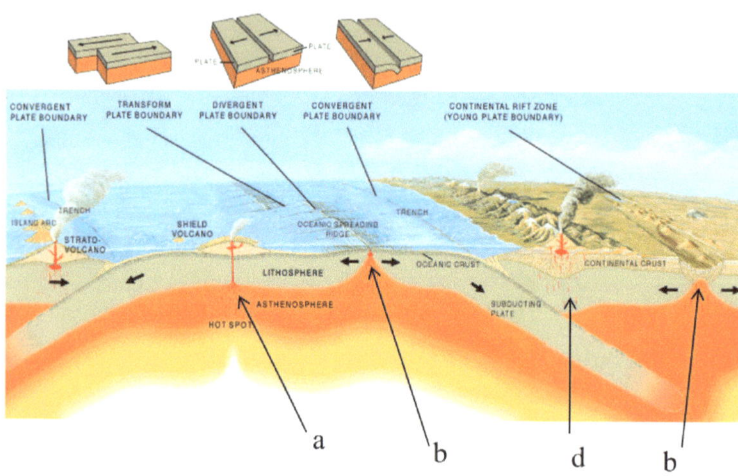

Figure 9.5 | Areas of magma generation at certain plate boundaries (b and d)/, or within a plate due to a hotspot (a). Arrows a, b, and d correspond to the same arrows as shown in Figure 9.1.

9.2.1 LAB EXERCISE

Part A Magma Generation and Plate Tectonics

Refer to Figure 9.4 to help answer the questions. The exercises that follow use Google Earth. For each question (or set of questions) paste the location that is given into the "Search" box. When finding your locations in Google Earth, be sure to zoom out to higher eye elevations in order to see all of the important features of each area.

1. Type 19 53 48.36 N 155 34 58.11 W in the search bar on Google Earth, and zoom out to an eye altitude of ~615 miles. The magma that resulted in the formation of these islands was generated by what process?
   a. decreased pressure (arrow "b")
   b. the addition of water (arrow "d", which shifts the solidus to the left)
   c. increased temperature (arrow "a")

2. Type 50 16 27.25 N 29 22 05.11 W in the search bar on Google Earth, and zoom out to an eye altitude of ~2663 miles. The magma generated at this location is due to:
   a. decreased pressure (arrow "b")
   b. the addition of water (arrow "d", which shifts the solidus to the left)
   c. increased temperature (arrow "a")
3. Type 36 11 46.85 S 71 09 48.03 W in the search bar on Google Earth, and zoom out to an eye altitude of ~ 2865 miles. The magma responsible for this volcano was generated by what process?
   a. decreased pressure (arrow "b")
   b.the addition of water (arrow "d", which shifts the solidus to the left)
   c. increased temperature (arrow "a")
4. Type "San Andreas Fault" in the search bar on Google Earth; no magma is generated here, because this plate boundary is_.
   a. a divergent plate boundary, with decompression melting occurs
   b. a convergent plate boundary, where water lowers the melting temperature of rock
   c. a transform plate boundary, where no magma is produced

9.3 MAGMATIC PROCESSES OCCURRING WITHIN THE EARTH'S CRUST

Once magma is generated by one of the mechanisms mentioned earlier (increased temperature, decreased pressure, or by adding water), the magma rises upward through the surrounding rock mainly through pre-existing fractures in the brittle lithosphere.

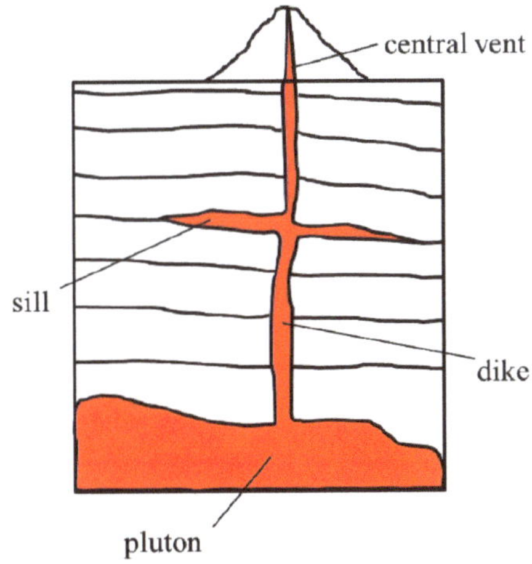

Figure 9.6 | Magma that crystallizes underground will form a pluton, an igneous rock body with a phaneritic texture. Magma can travel further upward through the crust by way of fractures through the rock layers to form a dike or magma may migrate in between rock layers and form a sill.

A lot of magma stops rising upward through the continental crust because it has encountered an area in the crust that has the same density as the magma. This is a natural occurrence when matter is under the influence of gravity: denser material will sink, whereas lower density material will rise upward (think of a lava lamp containing two liquids of different densities). This magma will cool slowly underground, and crystallize completely into an igneous rock body called a pluton (Figure 9.6). A lot of these plutonic rock bodies are found along our western coast of North America and other areas where magma must move upward through thick crust. Some of this magma may reach the surface either through a fissure or through a central vent in a volcanic structure; as the majority of magma stays underground and crystallizes into plutons, the smaller amount of magma that continues upward towards the surface may eventually crystallize completely within these fissures and become a **dike** or **sill** (Figure 9.6).

Dikes and sills can be found on the surface of the earth due to uplift and erosion of the surrounding rock layers; sills will be oriented parallel to the surrounding rock layers, but dikes cut across them and are therefore easy to identify on the Earth's surface.

Recall that magma is generated by partial melting of a source rock, and if we refer back to Bowen's Reaction Series (Igneous Rock chapter, or Figure 9.7 in this chapter), we see that minerals have different melting temperatures. If a rock undergoes partial melting, the minerals that melt are those minerals on the bottom of the Reaction Series that are rich in silica (SiO2) but poor in iron and magnesium. This means that magma generated by partial melting will be richer in silica and contain less iron and magnesium than the source rock (the rock that partially melted). The different magma types (and the solid rock forms that they make) in Figure 9.7 are listed in order of decreasing iron and magnesium, and increasing silica, with ultramafic rocks as the most iron and magnesium rich, and silica poor, and the felsic rocks as the most deficient in iron and magnesium but containing the most silica. Therefore partial melting of a mantle rock, which has an ultramafic composition, will produce a mafic magma; likewise, partial melting of a mafic rock can produce an intermediate magma type, and partial melting of an intermediate rock can produce felsic magma.

There are other ways that felsic and intermediate magmas are produced; for example, magma that is moving upward (ascending) through the crust may melt some of those crustal rocks, which can change the magma's composition. Recall that the continental crust is the lowest density rock type of the earth, of granitic composition, and is quite thick at convergent plate boundaries (see arrow "d" in Figure 9.5); at convergent plate boundaries, magma may have to travel through this thick portion of the continental crust. At any time before the magma is completely crystallized, it can melt the surrounding crust and perhaps end up with a more silica rich composition, therefore an intermediate or even a mafic magma will become more felsic as it travels through the continental crust. Whereas most magma crystallizes below ground, a portion of it does erupt onto the surface of the earth, either as a passive lava flow or under more explosive conditions such as the May 1980 eruption of Mount St. Helens, WA. Whether the style of eruption is passive or active (explosive) depends on the magma composition.

9.3.1 LAB EXERCISE

Part B Association of Magma Type with Tectonic Setting

The questions in this exercise demonstrate the control that tectonic setting has on the type of magma produced. For the Google Earth questions, copy and paste the latitude and longitude coordinates into the search bar (or just type them in).

5. A(n) ___________ at the Mid-Ocean Ridge, where oceanic plates are diverging and magma is generated by partial melting of the mantle.
   a. ultramafic magma is produced
   b. mafic magma is produced
   c. intermediate magma is produced
   d. felsic magma is produced
6. Type 19 28 19.70 N 155 35 31.94 W in the search bar on Google Earth, and zoom out to an eye altitude of ~119 miles. This volcano is composed of:
   a. mafic rocks, because a hotspot partially melted the mantle below the oceanic crust
   b. mafic rocks, because a hotspot partially melted the oceanic crust
   c. ultramafi rocks, because a hotspot partially melted the mantle below the oceanic crust
   d. ultramafic rocks, because a hotspot partially melted the oceanic crust
7. Type 35 35 08.45 S 70 45 08.22 W in the search bar in Google Earth. This volcano formed from an intermediate magma type, because:
   a. subduction of oceanic crust beneath the continental crust occurs here
   b. continental crust is subducting, causing magma to form
   c. a hot spot is partial melting the continental crust
   d. this is the result of a divergent plate boundary
8. Type 36 40 41.62 N 108 50 17.22 W in the search bar in Google Earth. The dark colored rock that forms a straight line on the surface is most likely:
   a. a pluton of ultramafic rock
   b. a dike of ultramafic rock
   c. a sill of mafic rock
   d. a dike of mafic rock

9.4 MAGMA COMPOSITION AND VISCOSITY

The chapter on igneous rocks explains that the classification of these rocks is somewhat determined by their mineral composition. Ultramafic rocks are igneous rocks predominantly consisting of olivine, with minor quantities of calcium-rich plagioclase and pyroxene, while quartz, muscovite, and potassium feldspar are characteristic minerals of felsic rocks (Figure 9.7). We must examine the mineral composition of these rocks, as igneous rocks are the crystallized product of solidified magma, and the minerals that develop during cooling are contingent upon the chemical makeup of the magma; for instance, mafic magma will yield a mafic rock rich in ferromagnesian minerals such as pyroxene and amphibole, while lacking quartz, which is consistently found in felsic rocks.

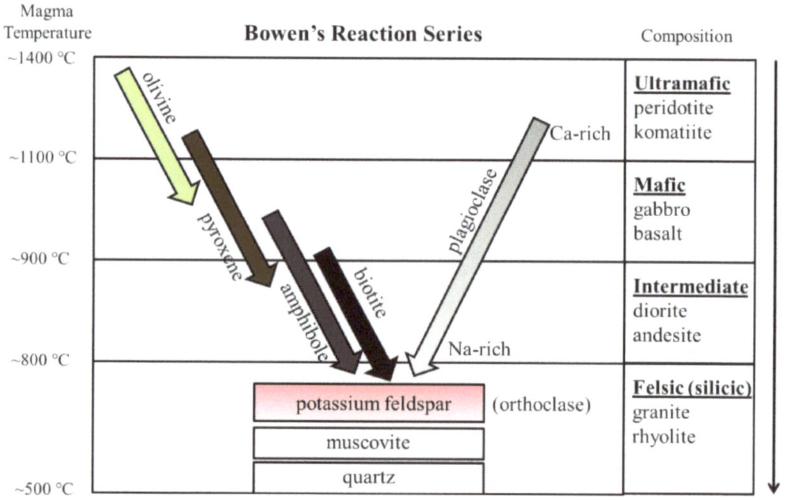

Figure 9.7 | Bowen's Reaction Series, and the magma compositions with the corresponding rock names listed on the right.

Most minerals in igneous rocks are silicate minerals, and all minerals depicted in Bowen's Reaction Series are part of the silicate mineral group. All silicate minerals possess crystal structures of silica tetrahedra (a silicon atom bonded to four oxygen atoms), which can be interconnected in various configurations to create sheets, linked chains, or a three-dimensional framework. Low-temperature minerals (quartz, muscovite, and orthoclase) possess a greater number of interconnected tetrahedra compared to high-temperature minerals (olivine and pyroxene). Figure 9.8A illustrates the linkage of silica tetrahedra in the formation of the material quartz. This is a two-dimensional representation of a three-dimensional structure, with numerous additional tetrahedra interconnected in the region above and beneath the typed page. The significance of this picture lies in the behavior of silica tetrahedra during quartz melting; picture 9.8B illustrates that while the crystalline structure is obliterated, the tetrahedral connections are preserved, though deformed. The connections connecting these tetrahedra are robust, and the temperatures of magma are insufficient to sever these bonds. This indicates that magmas capable of crystallizing quartz will possess numerous linked tetrahedra, while mafic magmas, lacking sufficient silicon for quartz crystallization, will instead form minerals with fewer linked tetrahedra.

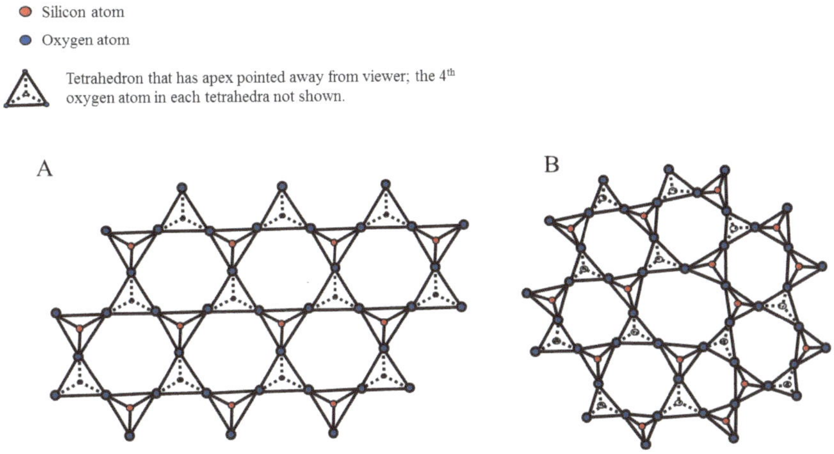

Figure 9.8 | (A) Regular repeated structure of silica tetrahedra in quartz; note that the 4th oxygen of each tetrahedra is not drawn for simpler viewing. These hexagonally arranged tetrahedral are joined to others to form a 3-dimensional framework. (B) shows how the tetrahedra stay linked when quartz melts; the arrangement of tetrahedral is irregular, but the links are maintained.

What is the significance of the silica concentration (the quantity of interconnected tetrahedra) in magma? A substantial quantity of interconnected silica tetrahedra will produce magma or lava with high viscosity, indicating a diminished capacity for fluidity. The temperature of lava influences its viscosity; consider the flow of ketchup from a refrigerator compared to that from a pantry; the colder ketchup exhibits more viscosity. In the context of magmas or lavas, higher temperatures facilitate flow, while reduced silica content correlates with decreased viscosity (refer to the right side of the diagram in Figure 9.7). Mafic lavas can flow more rapidly than intermediate or felsic lavas. The silica content of magma influences both the morphology of the volcano and the eruption style, determining whether the eruption will produce flowing lava or explosive magma.

9.4.1 LAB EXERCISE

Part C Magma Viscosity

The following questions address what factors control how fast a magma or lava can flow. The resistance to flow (viscosity) depends primarily on the magma or lava composition, and is also affected by temperature.

9. Intermediate lavas can flow….._than mafic lavas, due to the………viscosity.
   a. slower, higher b. slower, lower
   c. faster, higher d. faster, lower

10. Let's relate food items to magmas of different viscosity; if we compare how honey and water flow when poured from a container:
    a. then the honey represents felsic magma, and the water represents mafic magma
    b. then the honey represents mafic magma, and the water represents mafic magma.

11. Imagine putting the honey in the refrigerator overnight; will its viscosity be affected?
    a. yes; the viscosity will increase b. yes, the viscosity will decrease
    c. no. there will be no change as the composition stays the same

12. Keep the honey in mind while you answer this question: when it is first erupted, basalt lava typically erupts at 1200°C; after flowing away from the vent, the temperature falls, therefore the viscosity of the basaltic lava will:
    a. increase b. decrease c. stay the same

9.5 VOLCANIC TOPOGRAPHY AND ERUPTION MODALITIES

The size, shape, and eruptive style of any volcano ultimately depend on the magma composition. We will focus mainly on mafic and felsic magmas as intermediate magmas have properties that are intermediate between these two types, and ignore the ultramafic magma as this type no longer forms (due to a cooler earth's interior). As mentioned earlier, mafic magmas are lower in silica and are therefore characterized as having a low viscosity. As mafic magma erupts on to the surface through a central vent (see Figure 9.6, and photo in Figure 9.3), the magma (now called lava) will spread out quite easily due to its low viscosity. Mafic lava flows can travel quite far before solidifying completely, and the type of volcano that forms from mafic lava is called a shield volcano (Figure 9.9).

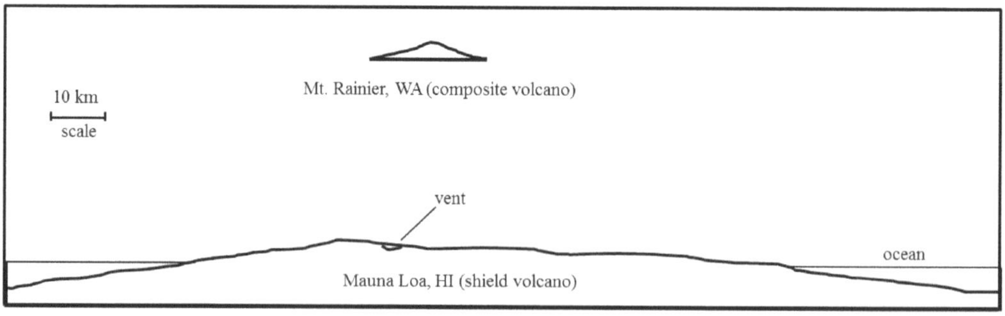

Figure 9.9 | A comparison of a typical shield volcano versus a smaller, steeply sloped composite volcano. The size and shape of these volcano types are due to the different types of magma and associated eruption styles.

Shield volcanoes are very broad at the base, and have relatively gentle slopes. There are several locations around the world where large amounts of mafic basalt flows are found, without forming a shield volcano: places such as the Deccan Traps in India, Siberian Traps in Russia, and here in the U.S. (Columbia River Basalt Group). Such vast outpourings of mafic lava are called flood basalts, and are believed to be caused by mantle plumes (hotspots), which partially melt the mantle beneath the earth's crust. The mafic magma that is generated by the mantle plume reaches the earth's surface through fractures (fissures) instead of one central vent.

Figure 9.9 also shows a much smaller volcano type, called a **composite volcano**. This volcano also forms from eruptions through a central vent, but the smaller size indicates that any lava that is generated from a central vent did not travel far before solidifying completely, which indicates a more viscous magma, such as an intermediate or felsic magma type. The term "composite" comes from the layers of lava flows, and the accumulation of ash and other volcanic material produced during a more explosive type of eruption due to the dissolved gases present in the magma (Figure 9.10, and photos in Figures 9.1 and 9.2).

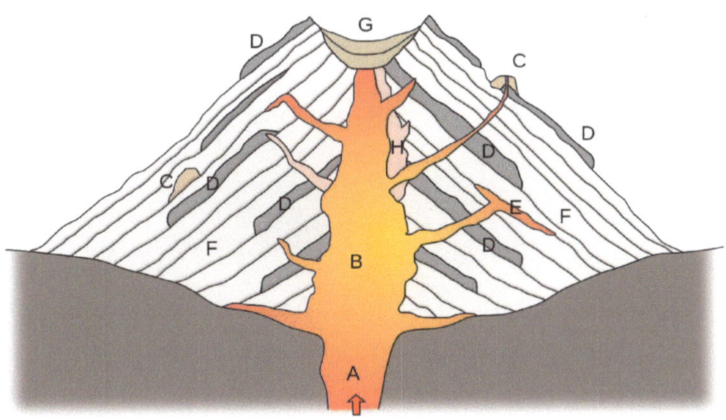

Figure 9.10 | Diagram of a composite volcano. Notice the structure is composed of layered ashflows (F) and viscous lava flows (D). Also of note is the buildup of viscous lava at locations labelled C; these are lava domes.

It is common to have a certain amount of dissolved gases within magma, and some gases may escape from the magma while still underground, but the most spectacular release of these gases occurs during eruption. The common gases associated with magma are usually water vapor, carbon dioxide, carbon monoxide, and hydrogen sulfide. Mafic magmas erupting onto the surface may form a lava fountain which spurts the magma (now called lava) into the air, where the height of the lava fountain depends on the gas content. However, gases within felsic magmas are not released as easily due the magma's high viscosity; as viscous magmas move closer to the earth's surface, the dissolved gases are under less pressure and large gas bubbles can form. At some point the gas pressure becomes so great that an explosive eruption occurs, and fragmented volcanic material is released. This material can travel great distances along air currents, but enough deposits around the central vent to build up the composite volcano structure. Once the gases are released and the explosive eruptions end, felsic magma may still be extruded onto the surface inside the volcanic crater. This felsic material, now a viscous lava, will flow with great difficulty, and a large amount of the felsic lava will cool completely in the area around the central vent, and may end up "plugging" the vent and forming what is called a **lava dome** (Figure 9.11).

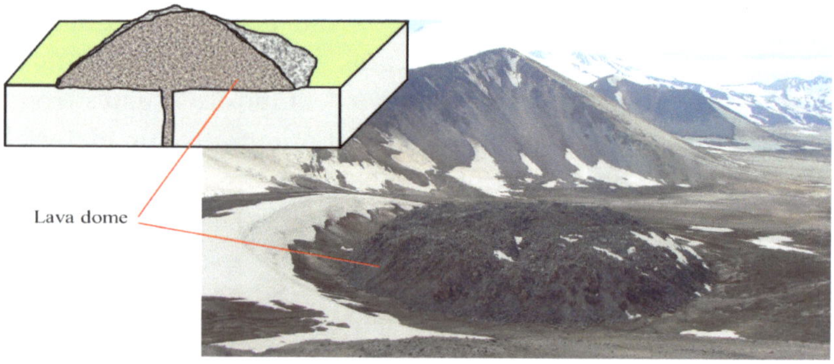

Figure 9.11 | Lava domes are extrusions of viscous lava on the surface of the Earth. The lava does not travel too far from the vent before solidifying completely.

9.5.1 LAB EXERCISE

Part D Volcanic Landforms

The composition of a magma or lava may also control what type of volcanic features or landforms are seen on the earth's surface. In this exercise, you will use Google Earth to identify these landforms. Figures 9.9 through 9.11 may help you in this section. For the Google Earth questions, copy and paste the latitude and longitude coordinates into the search bar (or just type them in).

13. Type 43 25 04.18N 113 31 37.38 W in the search bar of Google Earth. Zoom out to an eye elevation of ~90 miles. Based on the size of the area which is dark colored and sparsely vegetated, this region is:

    a. basaltic dike b. a shield volcano

    c. a lava dome d. a flood basalt

14. Click on the nearby photo icon (SW of your latitude/longitude coordinate in Question 13) to view a picture of the area as seen from the ground. This is a picture of:

    a. basalt flow b. a felsic dome

    c. a plutonic rock d. an ultramafic rock

15. How many miles in length is this dark colored feature? (measure the greatest length)
    a. ~8 miles
    b. ~20 miles
    c. ~40 miles
    d. ~50 miles
16. Type 46 12 07.84N 121 31 02.85W in the search bar of Google Earth. This volcano (Mt. Adams) has a lot of snow cover and small glaciers on it, but you can still see the volcanic rock, especially on the eastern flank (side) of the volcano. Zoom in to an eye altitude of ~9000ft to closely examine the rocks on this eastern side of the volcano; do you see any evidence of layering? This volcano is:
    a. shield volcano
    b. a lava dome
    c. a composite volcano
    d. a large volcanic dike
17. How tall is Mt. Adams, and what is the length of its base (in the widest, N-S dimension)? You will need to zoom out to an eye altitude of ~17 miles to measure the base.
    a. ~12,290 ft above sea level, and over 7 miles long
    b. ~11, 890 ft above sea level, and over 4 miles long
    c. at least 12, 703 ft above sea level, and over 6 miles long
    d. unable to determine because of the ice
18. Compared to the Big Island of Hawaii (19 53 48.36 N 155 34 58.11 W, or refer to Figure 9.6), which is actually the volcano Mauna Loa, Mt. Adams is:
    a. the same height, but the base is a lot smaller
    b. the same height, but the base is a lot wider
    c. smaller in height, and smaller at the base
    d. smaller in height, but larger at the base
19. Type in 58 15 58.56 N 155 09 35.98 W in the search bar of Google Earth. Examine the shape of this feature by zooming in to an eye altitude of ~3330 ft; use the eye icon in the upper right corner to rotate the view. Now zoom out to an eye altitude of ~7306 ft to see the entire structure and the surrounding area. Based on the size and appearance of this volcanic feature, this is a:
    a. dike
    b. shield volcano
    c. lava dome
    d. flood basalt

9.6 VOLCANIC HAZARDS

In comparing shield and composite volcanoes, it is evident that while shield volcanoes are more substantial, they pose significantly less hazard to the people than the tiny composite volcanoes. Shield volcanoes generate basaltic lavas that may erupt in fountains at the vent due to gas release, but thereafter flow passively down the volcano's slopes (Figures 9.3 and 9.12). Aside from property destruction, those residing on or near a shield volcano are unlikely to succumb to a volcanic explosion. Composite volcanoes exhibit a different phenomenon; their explosive eruptions generate substantial quantities of volcanic fragments, known as pyroclastic debris or tephra, which vary in size from fine dust and ash to sizable blocks (or bombs) of volcanic material (Figure 9.13). Initially, pyroclastic debris ascends into the atmosphere within an eruption column, subsequently dispersing in accordance with the prevailing wind direction. However, during the explosive phase of the eruption, the central vent expands as surrounding rocks are ejected; this enlargement of the vent diminishes the upward momentum of the eruption column. The pyroclastic material descends the volcano's slopes in a pyroclastic flow, a perilous amalgamation of heated volcanic substances and toxic gasses..

Figure 9.12 | Photo of a lava fountain, caused by gas release from a mafic magma. Notice the mafic lava flows down the sides of this structure.

Despite the high danger posed by pyroclastic flows, the majority of fatalities linked to composite volcanoes result from mudflows known as lahars. Numerous composite volcanoes are topped with snow and ice, and even a minor eruption can lead to meltwater cascading down the volcano's slopes. This water can rapidly erode the ash and other volcanic detritus on the slopes of the volcano, leading to a swiftly flowing slurry of mud and heavier debris, including trees and boulders. Lahars rapidly traverse river courses and pose a threat to any town or metropolis situated in the low-lying regions downstream from the volcano (Figure 9.13). Lahars may also be produced by substantial precipitation in the region. Numerous ancient mudflow deposits are identified in the Cascade region of Washington and Oregon. Due to the closeness of major cities and several settlements near these dormant volcanoes, as well as the recently designated "active" Mt. St. Helens, volcano monitoring systems have been established. Seismic activity is monitored because earthquakes result from the upward movement of magma beneath the volcanic structure, and GPS technology is employed to detect alterations in slope, as magma can expand the sides of the volcano and elevate the angle of its slope. The stream valleys on the volcano are also monitored to detect lahars, which can occur at any moment, irrespective of the active or dormant phase of a composite volcano.

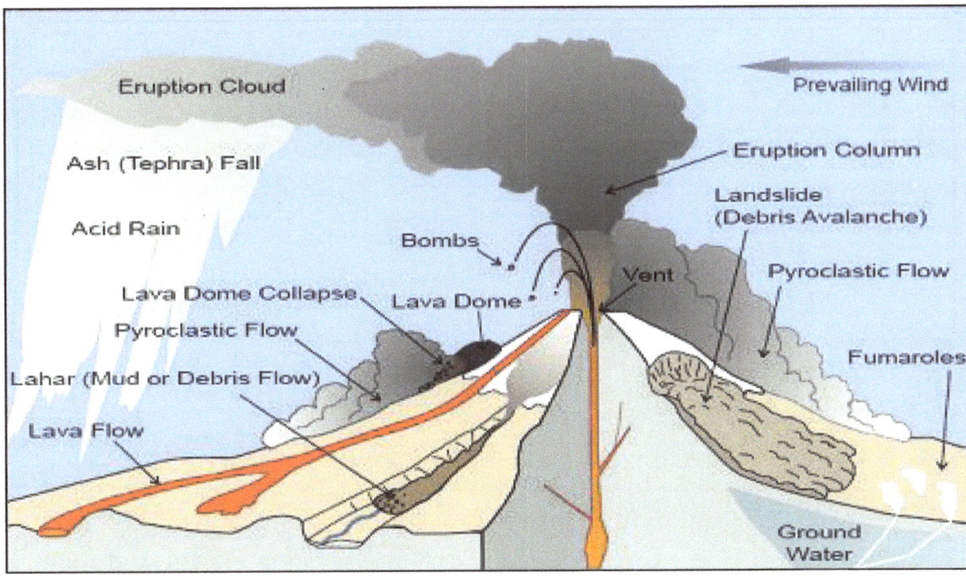

Figure 9.13 | Diagram demonstrating the hazards and other features associate/d with composite volcanoes.

9.6.1 LAB EXERCISE

Part E Volcanic Hazards

Potential hazards associated with certain volcanic types can be identified using topographic maps or aerial photographs. In this exercise, we will use Google Earth.

20. Type Mount St. Helens, WA in the search bar in Google Earth and examine the crater at an eye altitude of ~26,000 ft. Based on the appearance of the crater:
    a. a lahar removed the north side of the volcano
    b. a pyroclastic eruption removed the north side of the volcano
    c. a lava dome grew so large that it is higher than the north side of the crater
    d. a glacier has eroded the north side of the crater

21. Type Mount Rainier, WA in the search bar in Google Earth. The Carbon River flows from the north part of Mount Rainier (fed by meltwater from the Carbon Glacier on the flank of the volcano). At an eye altitude of ~20,000ft, follow the course of the Carbon River, past the town of Carbonado and stop at the town of Orting. Based on the locations of these two towns, which do you think is in danger from a lahar?
    a. Both are in danger of lahars
    b. Carbonado, because it is closer to Mount Rainier
    c. Orting, because it is in a low lying area along Carbon River
    d. Neither one is in danger because Mount Rainier is dormant

9.7 STUDENT RESPONSES

1. Type 19 53 48.36 N 155 34 58.11 W in the search bar on Google Earth, and zoom out to an eye altitude of ~615 miles. The magma that resulted in the formation of these islands was generated by what process?
   a. decreased pressure (arrow "b")
   b. the addition of water (arrow "d", which shifts the solidus to the left)
   c. increased temperature (arrow "a")
2. Type 50 16 27.25 N 29 22 05.11 W in the search bar on Google Earth, and zoom out to an eye altitude of ~2663 miles. The magma generated at this location is due to:
   a. decreased pressure (arrow "b")
   b. the addition of water (arrow "d", which shifts the solidus to the left)
   c. increased temperature (arrow "a")
3. Type 36 11 46.85 S 71 09 48.03 W in the search bar on Google Earth, and zoom out to an eye altitude of ~ 2865 miles. The magma responsible for this volcano was generated by what process?
   a. decreased pressure (arrow "b")
   b. the addition of water (arrow "d", which shifts the solidus to the left)
   c. increased temperature (arrow "a")
4. Type "San Andreas Fault" in the search bar on Google Earth; no magma is generated here, because this plate boundary is_.
   a. a divergent plate boundary, with decompression melting occurs
   b. a convergent plate boundary, where water lowers the melting temperature of rock
   c. a transform plate boundary, where no magma is produced
5. A(n) ______at the Mid-Ocean Ridge, where oceanic plates are diverging and magma is generated by partial melting of the mantle.
   a. ultramafic magma is produced
   b. mafic magma is produced
   c. intermediate magma is produced
   d. felsic magma is produced
6. Type 19 28 19.70 N 155 35 31.94 W in the search bar on Google Earth, and zoom out to an eye altitude of ~119 miles. This volcano is composed of:
   a. mafic rocks, because a hotspot partially melted the mantle below the oceanic crust
   b. mafic rocks, because a hotspot partially melted the oceanic crust
   c. ultramafi rocks, because a hotspot partially melted the mantle below the oceanic crust
   d. ultramafic rocks, because a hotspot partially melted the oceanic crust

7. Type 35 35 08.45 S 70 45 08.22 W in the search bar in Google Earth. This volcano formed from an intermediate magma type, because:

   a. subduction of oceanic crust beneath the continental crust occurs here

   b. continental crust is subducting, causing magma to form

   c. a hot spot is partial melting the continental crust

   d. this is the result of a divergent plate boundary

8. Type 36 40 41.62 N 108 50 17.22 W in the search bar in Google Earth. The dark colored rock that forms a straight line on the surface is most likely:

   a. a pluton of ultramafic rock　　b. a dike of ultramafic rock

   c. a sill of mafic rock　　d. a dike of mafic rock

9. Intermediate lavas can flow______than mafic lavas, due to the_____viscosity.

   a. slower, higher　　b. slower, lower

   c. faster, higher　　d. faster, lower

10. Let's relate food items to magmas of different viscosity; if we compare how honey and water flow when poured from a container:

    a. then the honey represents felsic magma, and the water represents mafic magma

    b. then the honey represents mafic magma, and the water represents mafic magma.

11. Imagine putting the honey in the refrigerator overnight; will its viscosity be affected?

    a. yes; the viscosity will increase

    b. yes, the viscosity will decrease

    c. no. there will be no change as the composition stays the same

12. Keep the honey in mind while you answer this question: when it is first erupted, basalt lava typically erupts at 1200°C; after flowing away from the vent, the temperature falls, therefore the viscosity of the basaltic lava will:

    a. increase　　b. decrease　　c. stay the same

13. Type 43 25 04.18N 113 31 37.38 W in the search bar of Google Earth. Zoom out to an eye elevation of ~90 miles. Based on the size of the area which is dark colored and sparsely vegetated, this region is:

    a. basaltic dike　　b. a shield volcano

    c. a lava dome　　d. a flood basalt

14. Click on the nearby photo icon (SW of your latitude/longitude coordinate in Question 13) to view a picture of the area as seen from the ground. This is a picture of:

    a. basalt flow　　b. a felsic dome

    c. a plutonic rock　　d. an ultramafic rock

15. How many miles in length is this dark colored feature? (measure the greatest length)
    a. ~8 miles
    b. ~20 miles
    c. ~40 miles
    d. ~50 miles

16. Type 46 12 07.84N 121 31 02.85W in the search bar of Google Earth. This volcano (Mt. Adams) has a lot of snow cover and small glaciers on it, but you can still see the volcanic rock, especially on the eastern flank (side) of the volcano. Zoom in to an eye altitude of ~9000ft to closely examine the rocks on this eastern side of the volcano; do you see any evidence of layering? This volcano is:
    a. shield volcano
    b. a lava dome
    c. a composite volcano
    d. a large volcanic dike

17. How tall is Mt. Adams, and what is the length of its base (in the widest, N-S dimension)? You will need to zoom out to an eye altitude of ~17 miles to measure the base.
    a. ~12,290 ft above sea level, and over 7 miles long
    b. ~11, 890 ft above sea level, and over 4 miles long
    c. at least 12, 703 ft above sea level, and over 6 miles long
    d. unable to determine because of the ice

18. Compared to the Big Island of Hawaii (19 53 48.36 N 155 34 58.11 W, or refer to Figure 9.6), which is actually the volcano Mauna Loa, Mt. Adams is:
    a. the same height, but the base is a lot smaller
    b. the same height, but the base is a lot wider
    c. smaller in height, and smaller at the base
    d. smaller in height, but larger at the base

19. Type in 58 15 58.56 N 155 09 35.98 W in the search bar of Google Earth. Examine the shape of this feature by zooming in to an eye altitude of ~3330 ft; use the eye icon in the upper right corner to rotate the view. Now zoom out to an eye altitude of ~7306 ft to see the entire structure and the surrounding area. Based on the size and appearance of this volcanic feature, this is a:
    a. dike
    b. shield volcano
    c. lava dome
    d. flood basalt

20. Type Mount St. Helens, WA in the search bar in Google Earth and examine the crater at an eye altitude of ~26,000 ft. Based on the appearance of the crater:
    a. a lahar removed the north side of the volcano
    b. a pyroclastic eruption removed the north side of the volcano
    c. a lava dome grew so large that it is higher than the north side of the crater
    d. a glacier has eroded the north side of the crater

21. Type Mount Rainier, WA in the search bar in Google Earth. The Carbon River flows from the north part of Mount Rainier (fed by meltwater from the Carbon Glacier on the flank of the volcano). At an eye altitude of ~20,000ft, follow the course of the Carbon River, past the town of Carbonado and stop at the town of Orting. Based on the locations of these two towns, which do you think is in danger from a lahar?

    a. Both are in danger of lahars

    b. Carbonado, because it is closer to Mount Rainier

    c. Orting, because it is in a low lying area along Carbon River

    d. Neither one is in danger because Mount Rainier is dormant

10 Chapter SEDIMENTARY ROCKS

10.1 INTRODUCTION:AN OVERVIEW

We are especially focused on the historical events and phenomena occurring on the Earth's surface due to their direct observability and their significant relevance to our lives and history. Sedimentary rocks serve as historical records, containing significant environmental indicators, biological remnants, and chemical signatures that provide insights into various topics, ranging from past catastrophes to biological production.

The classification of sedimentary rocks transcends mere nomenclature, as each designation encapsulates significant information about its origin, formation period, and the processes that contributed to its development. Each sedimentary rock is a riddle; by analyzing a collection of rocks, their stratification, the embedded fossils, and the patterns present, a geologist can reconstruct a complete environment and ecosystem. Resolving these issues serves as both an intellectual endeavor to enhance our comprehension of the surrounding world and a means to identify essential resources in our lives. Specifically, fossil fuels and various other natural resources are found within sedimentary rocks, including coal, natural gas, petroleum, salt, and the components used in wallboard and cement production. A comprehensive understanding of sedimentary rocks and their formation processes significantly impacts your daily life.

10.2 WEATHERING AND EROSION

Sedimentary rocks are created by the processes of weathering, erosion, deposition, and lithification of sediments. Sedimentary rocks are fundamentally formed of fragments from other rocks. This chapter will commence with a discussion of the process of weathering, which pertains to the breakdown of rocks. Weathering occurs in nature through two fundamental mechanisms. Initially, rocks can be physically fragmented into smaller constituents (for instance, striking a rock with a hammer), a process known as mechanical weathering. Alternatively, rocks can be disintegrated and transformed at the atomic level (similar to dissolving salt in water), a process known as chemical weathering. Various mechanisms exist for each type of weathering, resulting in significant variability in both the rate and manner of rock degradation, contingent upon the specific region and environmental conditions.

The predominant form of mechanical weathering involves the impact, fragmentation, and abrasion of rock caused by the flow of a fluid, such as water or air. The dimensions of the transported sediment are contingent upon the fluid type and the velocity of flow. A quick fluid, such as a swiftly moving river, can transport substantial particles and induce significant weathering, whereas a sluggish fluid, such a tranquil stream, exerts less weathering effects. The fluid's density regulates the size of particles that can be transported; for example, a denser fluid (such as water) can transport larger particles than a less dense fluid (such as air). A prevalent mechanism of mechanical weathering is known as frost wedging, which transpires when water infiltrates a fissure in the rock and subsequently freezes. Water possesses the distinctive characteristic of expanding upon freezing, exerting pressure on rocks and potentially fracturing boulders. The introduction and removal of heat or pressure can lead to the fracturing of rock; for example, pouring cold water on a heated light bulb may result in its shattering. This fracture may also occur in rocks when they cool rapidly or when substantial pressure is removed. Ultimately, flora, fauna, and humans can induce substantial weathering. The sediments subsequently experience erosion, the process of transporting material from its weathering location to its eventual deposition site, where it will consolidate into rock.

Rocks may undergo chemical weathering, mostly through one of three processes. The initial process, likely known to you, is termed dissolution. A mineral or rock is entirely disintegrated in water into discrete atoms or molecules. The individual ions can subsequently be transported with water and re-deposited as ion concentration increases, typically due to evaporation. Chemical weathering can alter the mineralogy and diminish the integrity of the original material, primarily due to water. A mineral can experience hydrolysis, a process in which a hydrogen atom from a water molecule substitutes for the cation in the mineral, typically transforming solids such as feldspar into a softer clay mineral. A mineral can undergo oxidation, a process in which oxygen atoms modify the valence state of a cation, typically occurring in metals and commonly referred to as rusting.

Chemical and mechanical weathering can together enhance the total rate of weathering. Chemical weathering diminishes the structural integrity of rocks, rendering them more susceptible to physical disintegration, whereas mechanical weathering amplifies the surface area of sediments, hence enhancing their exposure to chemical weathering. Consequently, settings exhibiting various forms of weathering might undergo rapid erosion. As you examine the subsequent sections regarding rocks and habitats, consider the forms of weathering necessary to produce the sediment that constitutes various sedimentary rocks, as well as the types of weathering anticipated in different environments..

10.3 DENTIFYING SEDIMENTARY ROCKS

The classification of sedimentary rocks is largely based on differentiating the processes that lead to their formation. The biggest division in types of sedimentary rocks types is based on the primary type of weathering that leads to the material building the sedimentary rock. If the rock is largely made from broken pieces (called clasts) of rock that have been mechanically weathered the rocks are referred to as Detrital or **Clastic Sedimentary Rocks.** Simply put, these are rocks that are composed of the broken pieces of other rocks. In this case, the mineralogy of the clasts is not important, but instead we need to note the properties of the sediment itself. Alternatively if the rock is largely the product of chemical weathering the classification is then based on the composition of the material as well as the processes involved in the materials precipitation from solution. **Chemical Sedimentary Rocks** form from the inorganic precipitation of minerals from a fluid. If the ions present within a fluid (water) become very concentrated either by the addition of more ions or the removal of water (by freezing or evaporation), then crystals begin to form. In this case the identification of the type of sedimentary rocks is based on the minerals present. If organisms facilitate the precipitation of these minerals from water we refer to the rocks as **Biochemical Sedimentary Rocks.** An example of biochemical precipitation is the formation of skeletal minerals in many organisms: from starfish and clams that grow calcite, to sponges that grow silica-based material, to humans that have bones made of hydroxyapatite. In many cases it is hard to differentiate whether a mineral was formed organically or inorganically, so in the current lab we will mostly group these two types of sedimentary rocks together. Rocks can also be formed from the carbon-based organic material produced by ancient life and are called **Organic Sedimentary Rocks.** Now we can discuss the identification and formation of particular sedimentary rocks.

10.3.1 Clastic Sedimentary Rocks

Weathering and erosion occur normally in areas that are at high elevation, such as mountains, while deposition occurs in lower areas such as valleys, lakes, or the ocean. The sediment is transported from the area of erosion to area of deposition by ice, water, or air. Not surprisingly, the sediment changes during its journey and we can recognize the amount of change and the distance the material has traveled, and the transport mechanism, by looking at its maturity (Figure 10.1). **Maturity** is defined as the texture and composition of a sedimentary rock resulting from varying amounts of erosion or sedimentary transport. Imagine a mountain composed of the igneous rock granite and let us explore how the sediment from this mountain changes as it makes the long distance trek via river to the ocean. The first process is just breaking the rock down into smaller pieces mechanically, which creates sediment that has large and small pieces, the pieces are jagged, and all of the minerals remain. The sizes of clasts in these rocks can range from large boulders, to cobbles, to pebbles, to the smallest particles, clay. As this sediment is transported in the river the pebbles collide with other pebbles and the rocks get smaller and the sharp edges are broken off. Also, as the slope of the land decreases the river slows leaving behind the large boulders and cobbles while carrying away the smaller particles. This results in sediments further from the source to be more uniform in size, which is a process called sorting. Chemical weathering also occurs, altering the feldspars into clay-sized particles. The end result of this process is the granite reduced from boulders and cobbles close to the mountain, to pebbles in the rivers, and finally to pure and uniform quartz sand at the beach and miniscule clay grains on the ocean floor. Therefore, different clastic rocks are found in different areas and have traveled different distances.

In this lab, we will look at three types of clastic rocks (Figure 10.1, Table 10.1), conglomerate, sandstone, and shale. Conglomerate is an immature sedimentary rock (rock that has been transported a short distance) that is a poorly sorted mixture of clay, sand, and rounded pebbles. The mineralogy of the sand and pebbles (also called clasts) can vary depending on its source. These rocks would be found on the continent in a several types of deposits such as ancient landslides or pebble beds in rivers. Sandstone is defined as a clastic sedimentary rock that consists of sand-sized clasts. These clasts can vary from jagged to rounded as well as containing many minerals or just quartz. Therefore, sandstone ranges from being relatively immature to mature which makes sense because we can find layers of sand associated with mountain rivers to pure white quartz beaches. Last we have shale, which is composed of clay particles and has a finely layered or fissile appearance. This extremely mature sedimentary rock is made from the smallest particles that can be carried by wind or barely moving water and can be found thousands of miles away from the original source.

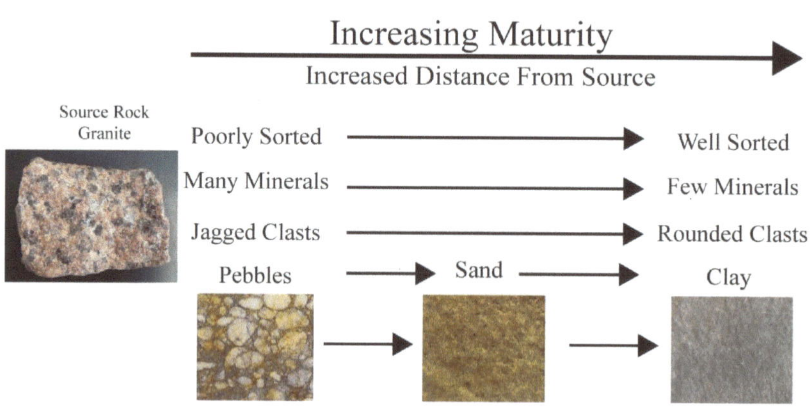

Figure 10.1 | Maturity in clastic sedimentary rocks showing how the sediments change as they are eroded further from their source.

Table 10.1 | Classification of Sedimentary rocks.

Detrital Sedimentary Rocks

Shale	Rock fragments smaller than 1/16 mm	Clay-sized particles that cannot be differentiated by the naked eye. May be fissile, splits into distinctive layers.	
Sandstone	Rock fragments between 1/16 - 2mm	Composed of sand-sized rock fragments. The fragments can vary in mineralogy including mainly quartz, along with feldspar, and clay	
Conglomerate	Rock fragments ranging in size with the largest >2mm	Poorly sorted mixture of rock fragments including subround or rounded pebbles.	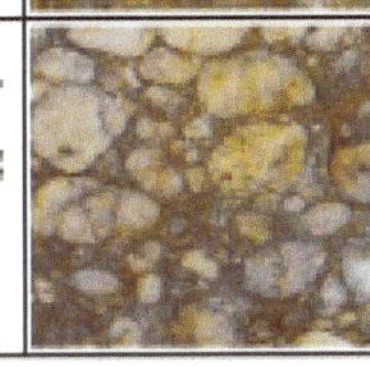

Chemical and Biochemical Sedimentary Rocks

Crystalline Limestone	Calcite crystals or microcrystalline calcite	Masses of large interlocking calcite crystals or microscopic crystals not visiable with the naked eye. Effervesces in dilute HCl.	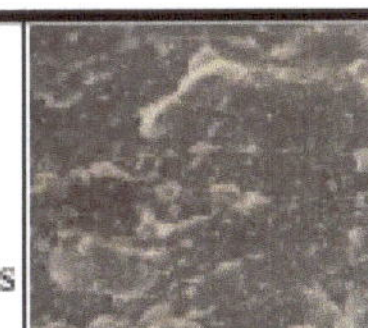
Coquina	Calcareous skeletal fragments of coral or shells	Consisting of gravel-sized shell or coral fragments composed of calcite or aragonite. Effervesces in dilute HCl.	
Dolostone	Microcrystalline dolomite	Masses of large interlocking dolomite crystals or microscopic crystals not visiable with the naked eye. Effervesces in dilute HCl.	
Chert	Microcrystalline polymorphs of quartz	Microcrystalline varieties of quartz. Scratches glass and may show conchoidal fracturing.	

Organic Sedimentary Rocks

Coal	Plant fragments and carbonized organic material	Black brittle rock with possible plant fragments. Light in weight with a sooty or shiny appearance. Dark gray streak.	

10.3.2 Biochemical and Chemical Sedimentary Rocks

As mentioned before, biochemical and chemical sedimentary rocks either precipitated directly from water or by organisms. The most recognizable chemical sedimentary rocks are evaporites. These are minerals that are formed by the precipitation of minerals from the evaporation of water. You have already examined multiple examples of these minerals/rocks in a previous lab, such as halite and gypsum. In this current lab, we will focus on siliceous and carbonate biochemical sedimentary rocks. Chert is a rock composed of microcrystalline varieties of quartz, and thus it has properties that are associated with quartz itself, such as conchoidal fracturing and hardness greater than glass. Chert is often formed deep in the ocean from silicious material that is either inorganic (silica clay) or biologic (skeletons of sponges and single-celled organisms) in origin. Carbonates are one of the most important groups of sedimentary rocks and as you have previously learned (Chapter 5), can result in distinctive landscapes (karst) and human hazards (sinkholes). Limestone is a sedimentary rock composed of the carbonate mineral calcite and can vary greatly in its appearance depending on how it is formed, but can easily be identified by its chemical weathering. Limestone composed of the mineral calcite undergoes dissolution in acids, in other words it effervesces dramatically when we apply dilute HCl. As with chert, limestone can be formed inorganically from a supersaturation of calcium and carbonate ions in water in varying environments from caves to tropical beaches. Limestone that consists of crystals of calcite or microcrystalline masses of calcite is called crystalline limestone. Alternatively, limestone can be formed biologically with the most striking example called a coquina, which are rocks made exclusively of fragmented carbonate (calcite or its polymorph aragonite) shells or coral. Lastly, we have dolostones, which are made from crystals (large or microscopic) of the mineral dolomite and are a carbonate that only weakly reacts to dilute HCl; you can scratch and powder dolostone to increase the surface area to see the reaction with acid. Dolostone is formed by the inorganic chemical alteration of limestones, therefore they are classified as chemical rather than biochemical sedimentary rocks.

10.3.3 Organic Sedimentary Rocks

Organic compounds are materials that contain a significant amount of the element carbon and are often associated with life. Organic sedimentary rocks are, therefore, rocks that consist mostly of carbon and are associated with significant biological activity. Other sedimentary rocks such as limestone and shale can contain carbon, but at much lower concentration (though shale can appear black from their carbon content). The most common organic sedimentary rock is coal, which is a very low density (light) black rock that has a dusty (sooty) or shiny appearance. It also produces a dark gray streak that can be seen both on a streak plate or a piece of paper. Coal is formed from the preservation and compaction of abundant plant material often in areas where oxygen is lacking, such as a swamp.

10.3.4 LAB EXERCISE

Materials

Unpack your HOL Sedimentary Rock bag with rocks labeled S1-S8 following what you see in Figure 10.2. In addition to these samples you will need the following from your HOL Rock Kit: 1) your glass plate, 2) your streak plate, 3) your hand lens, and 4) your bottle of diluted HCl.

Using Table 10.1, start identifying the rocks by separating out the organic sedimentary rocks from the chemical and biological sedimentary rocks from the clastic sedimentary rocks. Make sure to use all of the tools available including the glass plate and the diluted HCl to identify the chemical and biochemical sedimentary rocks (chert will be harder than glass, limestone will strongly react with dilute HCl, and dolostone will weakly react with dilute HCl when powdered). The streak plate can be helpful in identifying coal, which will easily produce a dark gray streak. Finally, use the hand lens to closely examine the size of the grains in the clastic sedimentary rocks. Once you are confident of your identifications, answer the following questions.

Figure 10.2 | The eight rocks (S1S8) in the Sedimentary Rocks Bag in the HOL kit.

Part A – Identifying Sedimentary Rocks

1. Sample S1 is called_.

a. Conglomerate	b. Crystalline Limestone
c. Coal	d. Shale
e. Coquina	f. Chert
g. Dolostone	h. Sandstone

2. Sample S1 is an example of a_______.
   a. Clastic Sedimentary Rock
   b. Organic Sedimentary Rock
   c. Chemical or Biochemical Sedimentary Rock
3. Sample S1 has the following characteristic:
   a. effervesces in diluted HCl acid
   b. weakly effervesces in diluted HCl acid if powdered
   c. contains fossil shells and effervesces in diluted HCl acid
   d. contains pebbles and finer sediments
   e. a sooty or shiny appearance
4. The formation of Sample S1 includes:_______.
   a. chemical weathering, transport of ions, precipitation of minerals, lithification
   b. mechanical weathering, transport of sediment a short distance, deposition of sediment, lithification
   c. photosynthesis, growth of organic material, deposition of organic materials, lithification
   d. chemical weathering, transport of ions, precipitation of minerals as shells by organisms, deposition, lithification
5. Sample S2 is called_.
   a. Conglomerate b. Crystalline Limestone
   c. Coal d. Shale
   e. Coquina f. Chert
   g. Dolostone h. Sandstone
6. Sample S2 is composed of__.
   a. clastic sediments b. calcite crystals c. dolomite crystals
   d. organic material e. calcite shells
7. Closely examine the individual grains in Sample S2. Which of the following is true about its maturity?
   a. It is mature because it contains a variety of different minerals.
   b. It is immature because it is poorly sorted.
   c. It is mature because it contains mostly rounded quartz grains.
   d. It is immature because the grains are jagged.
8. The formation of Sample S2 includes:_______.
   a. chemical weathering, transport of ions, precipitation of minerals, lithification

b. mechanical weathering, transport of sediment a long distance, deposition of sediment, lithification

c. mechanical weathering, transport of sediment a very short distance, deposition of sediment, lithification

d. photosynthesis, growth of organic material, deposition of organic materials, lithification

e. chemical weathering, transport of ions, precipitation of minerals as shells by organisms, deposition, lithification

9. Sample S3 is called_.

a. Conglomerate b. Crystalline Limestone

c. Coal d. Shale

e. Coquina f. Chert

g. Dolostone h. Sandstone

10. Sample S3 is an example of a_______.

a. Clastic Sedimentary Rock b. Organic Sedimentary Rock

c. Chemical Sedimentary Rock d. Biochemical Sedimentary Rock

11. Sample S3 has the following characteristic:

a. effervesces in diluted HCl acid

b.weakly effervesces in diluted HCl acid if powdered

c. contains fossil shells and effervesces in diluted HCl acid

d. contains pebbles and finer sediments

e. has a sooty or shiny appearance

12. The formation of Sample S3 includes:_______.

a. chemical weathering, transport of ions, precipitation of minerals, lithification

b. mechanical weathering, transport of sediment a short distance, deposition of sediment, lithification

c. chemical weathering, transport of ions, precipitation of minerals, lithification, chemical alteration

d. photosynthesis, growth of organic material, deposition of organic materials, lithification

e. chemical weathering, transport of ions, precipitation of minerals as shells by organisms, deposition, lithification

13. Sample S4 is called_.

a. Conglomerate b. Crystalline Limestone

c. Coal d. Shale

e. Coquina
f. Chert
g. Dolostone
h. Sandstone

14. Sample S4 is an example of a_______.
    a. Clastic Sedimentary Rock
    b. Organic Sedimentary Rock
    c. Chemical Sedimentary Rock
    d. Biochemical Sedimentary Rock
15. Closely examine the individual grains in Sample S4. Which of the following is true about its maturity?
    a. It is immature because it is poorly sorted.
    b. It is mature because it is poorly sorted.
    c. It is mature because it contains mostly rounded grains.
    d. It is immature because it contains mostly rounded grains.
    e. It is immature because it has clay-sized particles.
16. The formation of Sample S4 includes:_______.
    a. chemical weathering, transportation of ions, precipitation of minerals, lithification
    b. mechanical weathering, transportation of sediment a short distance, deposition of sediments, lithification
    c. photosynthesis, growth of organic material, deposition of organic materials, lithification
    d. chemical weathering, transportation of ions, precipitation of minerals as shells by organisms, deposition, lithification
17. Sample S5 is called .
    a. Conglomerate
    b. Crystalline Limestone
    c. Coal
    d. Shale
    e. Coquina
    f. Chert
    g. Dolostone
    h. Sandstone
18. Sample S5 is an example of a .
    a. Clastic Sedimentary Rock
    b. Organic Sedimentary Rock
    c. Chemical or Biochemical Sedimentary Rock
19. The history of formation of Sample S5 includes: .
    a. chemical weathering, transportation of ions, precipitation of minerals, lithification
    b. mechanical weathering, transportation of sediment a short distance, deposition of sediments, lithification
    c. photosynthesis, growth of organic material, deposition of organic materials, lithification

d. chemical weathering, transportation of ions, precipitation of minerals as shells by organisms, deposition, lithification

20. Sample S5 has the following characteristic:
    a. effervesces in diluted HCl acid
    b. weakly effervesces in diluted HCl acid if powdered
    c. contains fossil shells and effervesces in diluted HCl acid
    d. contains pebbles and finer sediments
    e. a sooty or shiny appearance
21. Sample S6 is called_.
    a. Conglomerate
    b. Crystalline Limestone
    c. Coal
    d. Shale
    e. Coquina
    f. Chert
    g. Dolostone
    h. Sandstone
22. The history of formation of Sample S6 includes:_____.
    a. chemical weathering, transportation of ions, precipitation of minerals either biologically or inorganically, lithification
    b. mechanical weathering, transportation of sediments a short distance, deposition of sediments, lithification
    c. photosynthesis, growth of organic material, deposition of organic materials, lithification
    d. chemical weathering, transportation of ions, precipitation of minerals as shells by organisms, deposition, lithification
23. Sample S6 is composed of__.
    a. clastic sediments
    b. microcrystalline calcite crystals
    c. microcrystalline dolomite crystals
    d. microcrystalline quartz crystals
    e. organic material
24. Sample S6 can be easily recognized by which of the following properties?
    a. conchoidal fracturing
    b. weakly effervesces in diluted HCl acid if powdered
    c. fissule appearance
    d. a sandpaper texture
    e. a sooty or shiny appearance

25. Sample S7 is called_.
    a. Conglomerate
    b. Crystalline Limestone
    c. Coal
    d. Shale
    e. Coquina
    f. Chert
    g. Dolostone
    h. Sandstone

26. Compared to Sample S2, how mature is Sample S7?
    a. more mature
    b. less mature
    c. same level of maturity

27. Sample S7 consists of______.
    a. fragments of calcite shells
    b. clay-sized sediments
    c. sand-sized sediments
    d. organic material
    e. dolomite crystals

28. Sample S7 can be easily recognized by which of the following properties?
    a. conchoidal fracturing
    b. weakly effervesces in diluted HCl acid if powdered
    c. fissile appearance
    d. a sandpaper texture
    e. a sooty or shiny appearance

29. Sample S8 is called_.
    a. Conglomerate
    b. Crystalline Limestone
    c. Coal
    d. Shale
    e. Coquina
    f. Chert
    g. Dolostone
    h. Sandstone

30. Sample S8 is an example of a_______.
    a. Clastic Sedimentary Rock
    b. Organic Sedimentary Rock
    c. Chemical Sedimentary Rock
    d. Biochemical Sedimentary Rock

31. Sample S8 is most likely to weather by which of the following processes?
    a. Dissolution
    b. Frost Wedging
    c. Oxidation
    d. Hydrolysis
    e. The addition and subtraction of heat

32. The history of formation of Sample S8 includes:_____.
    a. chemical weathering, transportation of ions, precipitation of minerals, lithification
    b. mechanical weathering, transportation of sediments a long distance, deposition of sediments, lithification

c. mechanical weathering, transportation of sediments a very short distance, deposition of sediments, lithification

d. photosynthesis, growth of organic material, deposition of organic materials, lithifilation

e. chemical weathering, transportation of ions, precipitation of minerals as shells by organisms, deposition, lithification

10.4 SEDIMENTARY FORMATIONS

Sedimentary rocks often show distinctive patterns that are unrelated to their type of rock, yet reflect events or conditions during deposition and are called **sedimentary structures.** These patterns in the rocks can be very informative to geologists attempting to reconstruct the environment in which a sedimentary rock was formed. Imagine wind blowing steadily along a beach; this wind pushes the sand into dunes that can be preserved in the rock record, informing us about the strength and direction of the wind along with the rock type.

Examples of sedimentary structures are given in Table 10.2, but let us discuss them in more detail. The sedimentary structures that most students are familiar with are ripples and dunes. We are familiar with seeing dunes at the beach or in deserts or smaller ripples in mud puddles. In each case, these ripples are formed from either a wind or water current.

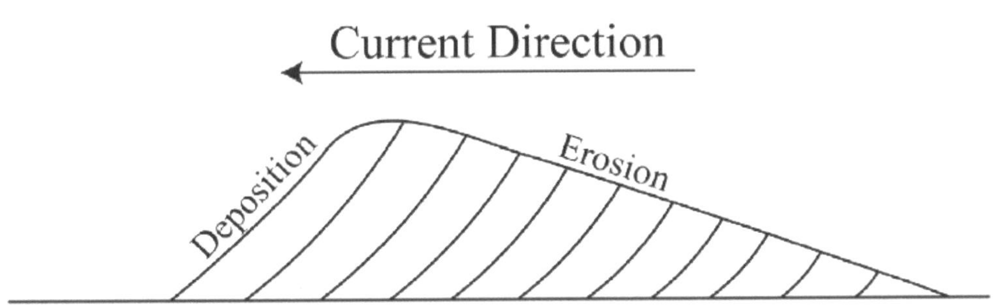

Figure 10.3 | Cross-section of a ripple showing the area of erosion and deposition.

Ripples form by the current pushing sediment into a pile; on the down-current side the sediment is shadowed and protected from 54the wind or water current. This means that there is erosion on the up-current side making a shallow slope and deposition on the steeper, down-current side. Therefore, if you cut a ripple in half and look at it in prospect you can see inclined layers of sediment building up on the steep down current side of the ripple. We can often see multiple layers of beds consisting of these inclined layers, which represent multiple generations of migrating ripples or dunes that are called cross-beds. Both ripples and cross-beds can indicate the presence and direction of the current in an environment.

Table 10.2 | Descriptions of common sedimentary structures. Images provided by (from top to bottom) Grand Canyon National Park, Zion National Park, B. Deline, James St. John, Grand Canyon National Park, and B. Deline.

Sedimentary Structure	Description	Required conditions	
Ripple Marks	Either symmetrical or asymetrical ripples present on bedding surfaces.	Water or air currents. Asymetrical ripples indicate unidirectional currents with the steep slope facing down current and the shallow slope facing up current. Symmetrical ripples indicate bidirectional currents.	
Cross-bedding	Layers of inclined beds often altering directions from bed to bed.	Water or air currents.	
Graded Beds	Beds of sedimentary rocks with a change in sediment size through the layer with smaller grains at the top of the bed and larger grains on the bottom.	A turbulent water current carrying sediment loses energy and slows.	
Raindrop impressions	Small indentations on a bedding plane from the impact of raindrops.	Soft muddy substrate that is exposed and subsequently buried with sediment.	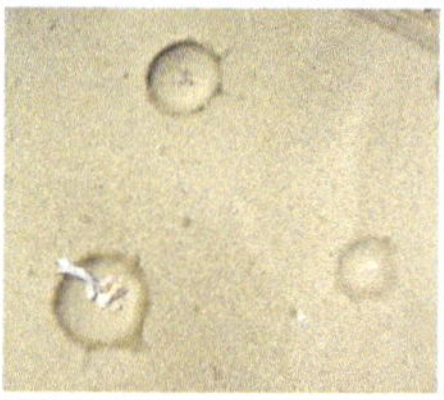
Mudcrack casts	Muddy cast infills of a polygonal pattern of cracks formed in mud as it drys.	Alternating wet and dry conditions create cracks in mud, which are then buried by sediment and preserved.	
Trace fossils	Tracks, burrows, or other traces left by the activity of plants or animals.	The presence of animals or plants. These traces occur where the organism lives and more specific information can be obtained based on the identity of the tracemaker.	

We can also see structures that indicate changes in the strength of a current. Imagine a fast moving current carrying a variety of sediment sizes; if the current slows it will no longer be able to carry the largest particles and they will be deposited first.

Then as the current continues to slow progressively smaller particles are deposited on top of the bigger particles, forming a sedimentary deposit called a graded bed. This graded bed is a sedimentary layer with larger clasts on the bottom and smaller clasts on the top. These types of beds can be the result of floods in a river or storms in the ocean. We can also observe features that are pretty self-explanatory such as casts of mud cracks (covered and preserved cracks that are the result of the drying of wet mud), and rain drop impressions (covered and preserved impacts of rain drops in soft mud). In both of these cases, the sedimentary structure tells us about the sediment, the water content, and its exposure at the surface above water level.

Related to sedimentary structures are trace fossils, which are patterns in the rocks that are caused by the activity of organisms. These can occur in many different ways and can indicate many different aspects about the environment depending on the trace and the identity of the tracemaker. Traces can be terrestrial such as footprints, burrows or dens, or the traces of roots, which can inform us about the climate, ecosystem, and the development of soils. Traces can also be found in freshwater and marine environments, such as burrows, borings, footprints, or feeding traces, which tell us about the sediment, chemistry of the poor water, and the life that lived within it.

10.5 DEPOSITIONAL ENVIRONMENTS

A depositional environment is the accumulation of chemical, biological, and physical properties and processes associated with the deposition of sediments that lead to a distinctive suite of sedimentary rocks. Sedimentary environments are interpreted by geologists based on clues within such as rock types, sedimentary structures, trace fossils, and fossils. We can then compare these clues to modern environments to reconstruct ancient environments. We can break up the numerous depositional environments found on earth into common environments we find on land **(continental environments)**, in the ocean **(marine environments)**, and at the interface between the two **(transitional environments)**. As each of the environments is briefly described, think about and fill in Table 3 with the following information: 1) select the image that best depicts the depositional environment in Figure 10.4 and put the corresponding letter (from Figure 10.4) in the first column, 2) the types of sediments and sedimentary rocks most likely to occur in that environment, 3) the maturity of those sediments if clastic, and 4) the types of sedimentary structures that may be present.

10.5.1 Continental Environments

There are many different environments on the continents, but again we are limited to those that are dominated by the deposition rather than the erosion of sediments. Erosion occurs in high altitude areas and although continents are overall topographically elevated compared to the oceans, there are several different areas on the continent where we get distinctive depositional properties. Continental depositional environments are dominated by clastic sedimentary rocks, largely because of their proximity to the source of the sediments.

Glacial depositional environments are controlled mostly by the weathering and erosion by glaciers and glacial meltwater. Glaciers most commonly occur in areas that are both high elevation and/or high latitudes. Glaciers are fairly slow moving (centimeters a day) and normally travel short distances from their source, but they can cause immense mechanical weathering. Glaciers grind and bulldoze rock and create piles of poorly sorted sediment called moraines. Glaciers also produce a significant though fluctuating amount of melt water, which flows through the moraines building a system of braided rivers. These rivers carry the small sediments further from the end of the glaciers into an area called the outwash plain, which consists of poorly sorted sediment.

We have spent a significant amount of time in chapter five discussing rivers and how they erode, transport, and deposit sediment. Sediments that are deposited through the action of rivers are referred to as **fluvial depositional environments.** The gradient and discharge of a river can greatly control the shape of the river, how it flows, and how it deposits sediment. Rivers alter sediment both chemically and physically as well as sort the sediment since there is a limit to the size or particles a river can carry. Within a meandering river we see several different types of sediments, from the pebbles and stones within the river channel to sandy point bars along the outer edge of the meander where the water slows. In addition, we see multiple types of sedimentary structures related to the flow of the river as well as those related to flood events.

We also have sediments deposited within lakes, which are called **lacustrine depositional environments**. Unlike rivers, lakes do not have rapidly flow water and thus there is significance less movement of sediment. The sediment that accumulates in lakes can come from several sources including rainwater carrying sediment into the lake from the shores, rivers that flow into the lake, and sediment that is transported by the wind. Once the sediment reaches the lake it remains undisturbed so we see thin layers of fine sediment, with varying amounts of trace fossil.

Lastly, there are **eolian depositional environment**, which are dominated by currents of wind rather than water. Since air is less dense than water, only smaller particles can be transported. In addition, wind is not restrained within distinctive channels like water and, therefore, the features of eolian deposits are more widespread than those of fluvial deposits. Certain areas have predominant wind patterns, such that the wind is fairly consistent in direction and strength, which can generate significant sedimentary structures. When water is present within these environments it is often in the form of seasonal lakes that undergo significant evaporation and sometimes leaving behind chemical sedimentary rocks (evaporite deposits).

10.5.2 Transitional Environments

The interface between the continents and oceans are complicated areas that can be influenced by rivers, ocean currents, winds, waves, and tides. In addition, the sediments that are present in these areas are a mixture of materials derived from the continents (clastic and organic) and those from the ocean (chemical and biochemical). Finally, with the abundant currents (both air and water), sediments influenced by the elements, and abundant life these areas, which results in abundant sedimentary structures and trace fossils. However, we can distinguish transitional areas that are dominated by different forces such as tides, ocean currents, and rivers.

Shorelines that are influenced by strong daily tidal currents are called **tidal mudflat depositional environments.** Tides are currents that are the result of the gravitational forces exerted by the moon and the rotation of the earth. Shorelines that have strong tidal currents as well as seafloors with low gradients can have large areas that are submerged during high tide and exposed to air during low tide. These areas often have smaller particles than a normal shoreline since the tidal currents can pull marine sediments into the area. In addition, the strong bidirectional currents, daily drying out, exposure to the elements, and abundant life create abundant indicators of these environments.

Shorelines that are dominated by ocean currents are called **beach depositional environments.** Shorelines have constant wind on and off shore that are the result of the difference in the way the land and water heat and cool through the day. These winds produce the waves that are iconic at the beach, but as these waves move onto shore they curve, mimicking the shape of the shore and result in a current that runs parallel to the shore itself. This current carries and deposits sand along the beach. In addition, the wind can also produce dunes, which promote a diverse and complicated ecosystem. Within beach depositional environments there are areas where rivers flow into the ocean, which are called **Deltas.** As a river that is carrying material empties into the ocean the water slows and deposits sediment. Most of the sediment is deposited at the mouth of the river with some spilling out into the surrounding areas building a distinctive fan of sediment. Since the sediment is coming from the river, the delta is largely a thick sequence of clastic sediment showing indications of the strong flow of the river.

10.5.3 Marine Environments

Marine depositional environments diff in multiple ways, but the controlling factors in the rocks that are produced is related to the proximity and supply of continental sediment, the water depth, and the community of organisms that live in the area. The further an environment is from the shore the less clastic sediment will be present and the area will have a higher concentration of the chemical and biological sedimentary rocks that are formed within the ocean. In addition, some organisms in the right environmental conditions can produce huge amounts of skeletal material.

Shallow marine depositional environments are areas that are close to shore, but always submerged. These areas have a significant amount of mature clastic sediment along with marine algae (like sea grass) as well as skeletal material from animals like coral, echinoderms (sea urchins and sand dollars), and mollusks (clams and snails). These areas can have a significant difference in their energy level from very shallow areas influenced by waves to deeper areas only influenced by large storms. A better understanding of the relative depth can often be determined based on the sedimentary structures as well as the community of organisms and types of trace fossils, which can be very sensitive to depth.

In warm tropical shallow water area we often find **reef depositional environments.** Reefs are formed through the growth of coral colonies building a large three-dimensional structure built from calcite skeletons. Corals can grow in many diff marine environments, but they can only produce reefs when their symbiotic algae that live within their tentacles can photosynthesize effectively, resulting in more energy for the coral to grow faster. Reefs also create a barrier between shallow water environments and ocean currents producing shallow, low-energy environment called lagoons. Lagoons can have thin layers of fine sediment that we would expect in quiet water along with chemical sedimentary deposits that are the result of evaporation.

Most of the ocean consists of **deep marine depositional environments.** These areas are beyond the reach of most clastic sediment other than the dust carried by the wind. Therefore, the sediment is being produced chemically and biologically within the ocean. The largest source of sediment in these deep settings is skeletal material from some of the smallest marine organisms. Multiple types of single-celled organisms can produce shells composed of either silica or calcite. These shells are mostly produced in the surface waters that are bathed in sunlight permitting photosynthesis. When the organism dies, these shells then rain down into deeper water; this slow accumulation of sediment produces fine layers of biochemical sedimentary rocks. In some cases, these shells are dissolved or altered before they reach the bottom (which can be miles away) and are precipitated as chemical sedimentary rocks. Obviously, there is not a clear boundary between shallow and deep-water environments given the gradient of the ocean floor. The deep marine depositional environment is normally thousands of feet deep and beyond the influence of even large storms.

10.5.4 Depositional Environments

Based on the descriptions of the different sedimentary environments along with your understanding of weathering, sedimentary rocks, and sedimentary structures please fill out the following table, which can then be used as a guide for answering the following questions on depositional environments.

Figure 10.4 | Modern examples of the discussed depositional

Table 10.3

Example from Figure 4	Depositional Environment	Description	Clastic Rock Maturity, distance from source, and type	Other sedimentary Rocks	Sedimentary Structures
Continental					
	Fluvial				
	Lacustrine				
	Glacial				
	Eolian				
Transitional					
	Beach				
	Mudflats				
Marine					
	Shallow Marine				
	Reefs				
	Deep Marine				

10.5.5 LAB EXERCISE

Part B – Depositional Environments

The following exercises use Google Earth to explore the depositional environments that are the source for sedimentary rocks. For each locality think about the types of sediments that are accumulating, the types of weathering that would occur, as well as the presence of any sedimentary structures.

33. Search for 39 05 52.46N 84 30 56.16E and zoom out to an eye altitude of ~30,000 ft. The large-scale structures in this sedimentary environment are asymmetrical ripples (known as dunes at this size). Zoom in and out and examine the sedimentary environment. What type of sedimentary environment is this?

    a. Lacustrine b. Fluvial c. Eolian d. Glacial

34. What type of weathering do you think is most prominent in this sedimentary environment?

    a. Mechanical weathering from a current

    b. Dissolution

    c. Frost Wedging

    d. Hydrolysis

35. Study the large dunes in this image (zoom out to an eye altitude of ~25,000 feet). These structures can indicate the direction that the wind is blowing. What is the predominant wind direction in this area? (Hint: it is easier to see these features if we exaggerate the vertical scale to do this go to Tools, options, and on the 3D view tab change the Elevation Exaggeration to 3. To do this on a Mac go to Google Earth then Preferences)

    a. north to south b. south to north c. east to west d. west to east

36. Search for 64 01 03.61N 16 52 56.63W and zoom out to an eye altitude of ~25,000 ft. What type of sedimentary environment is this?

    a. Lacustrine b. Fluvial c. Eolian d. Glacial

37. What type of weathering do you think is most prominent in this sedimentary environment? (Make sure you evaluate all choices.)

    a. Mechanical weathering from ice

    b. Dissolution

    c. Frost Wedging

    d. Hydrolysis

    e. Both a and c are correct

38. When the sediment you see today lithifies, what type of sedimentary rock would you expect to be most abundant in this area and how mature is this rock type?

    a. Shale, Mature b. Shale, Immature

c. Sandstone, Immature
d. Sandstone, Mature
e. Conglomerate, Immature
f. Conglomerate, Mature

39. Search for 44 42 16.98N 1 05 23.88W and zoom out to an eye altitude of ~17 miles. Notice the depth of this shallow bay (Arcachon Bay, France). This bay dries out at times and later greatly expands in size daily. What sedimentary structure would you likely find in this area because of this?
    a. graded beds
    b. finely layered beds
    c. cross beds
    d. mud cracks
40. What type of sedimentary environment is this?
    a. lacustrine
    b. tidal mudflats
    c. deep marine
    d. beach
41. Search for 20 20 23.94S 150 38 29.14E and zoom out to an eye altitude of ~25 miles (also zoom far out to notice where you are in the world). What type of sedimentary environment is this?
    a. shallow marine
    b. deep marine
    c. delta
    d. reef
42. Think about the origin of this marine formation and consider the latitude to assess climatic conditions. What is the most abundant type of rock you would expect to form in this sedimentary environment?
    a. Limestone
    b. Sandstone
    c. Coal
    d. Shale
43. If the sea level dropped 1,000 feet and this sedimentary environment stopped being built and began to break down, what type of weathering would be most likely to occur on these rocks?
    a. Fracturing from the addition or subtraction of pressure
    b. Dissolution
    c. Frost Wedging
    d. Hydrolysis
44. Search for 31 00 43.61N 81 25 40.92W and zoom out to an eye altitude of 1,500 ft. What type of sedimentary environment is this?
    a. shallow marine
    b. reef
    c. fluvial
    d. each
    e. tidal mudflat
45. What type of sediment would you expect to find in this depositional environment?
    a. Mostly mature clastic sediments with some biochemical and chemical sedimentary sediments

b. Mostly biochemical and chemical sedimentary sediments with some mature clastic sediments

c. Exclusively organic sediments

d. Exclusively biochemical sediments

e. Exclusively immature clastic sediments

10.6 STUDENT RESPONSES

1. Sample S1 is called_.

   a. Conglomerate b. Crystalline Limestone
   c. Coal d. Shale
   e. Coquina f. Chert
   g. Dolostone h. Sandstone

2. Sample S1 is an example of a_______.

   a. Clastic Sedimentary Rock
   b. Organic Sedimentary Rock
   c. Chemical or Biochemical Sedimentary Rock

3. Sample S1 has the following characteristic:

   a. effervesces in diluted HCl acid
   b. weakly effervesces in diluted HCl acid if powdered
   c. contains fossil shells and effervesces in diluted HCl acid
   d. contains pebbles and finer sediments
   e. a sooty or shiny appearance

4. The formation of Sample S1 includes:_______.

   a. chemical weathering, transport of ions, precipitation of minerals, lithification
   b. mechanical weathering, transport of sediment a short distance, deposition of sediment, lithification
   c. photosynthesis, growth of organic material, deposition of organic materials, lithifi ation
   d. chemical weathering, transport of ions, precipitation of minerals as shells by organisms, deposition, lithification

5. Sample S2 is called_.

   a. Conglomerate b. Crystalline Limestone
   c. Coal d. Shale
   e. Coquina f. Chert
   g. Dolostone h. Sandstone

6. Sample S2 is composed of__.

a. clastic sediments b. calcite crystals c. dolomite crystals

d. organic material e. calcite shells

7. Closely examine the individual grains in Sample S2. Which of the following is true about its maturity?

a. It is mature because it contains a variety of different minerals.

b. It is immature because it is poorly sorted.

c. It is mature because it contains mostly rounded quartz grains.

d. It is immature because the grains are jagged.

8. The formation of Sample S2 includes:_______.

a. chemical weathering, transport of ions, precipitation of minerals, lithification

b. mechanical weathering, transport of sediment a long distance, deposition of sediment, lithification

c. mechanical weathering, transport of sediment a very short distance, deposition of sediment, lithification

d. photosynthesis, growth of organic material, deposition of organic materials, lithifi ation

e. chemical weathering, transport of ions, precipitation of minerals as shells by organisms, deposition, lithification

9. Sample S3 is called_.

a. Conglomerate b. Crystalline Limestone

c. Coal d. Shale

e. Coquina f. Chert

g. Dolostone h. Sandstone

10. Sample S3 is an example of a_______.

a. Clastic Sedimentary Rock b. Organic Sedimentary Rock

c. Chemical Sedimentary Rock d. Biochemical Sedimentary Rock

11. Sample S3 has the following characteristic:

a. effervesces in diluted HCl acid

b. weakly effervesces in diluted HCl acid if powdered

c. contains fossil shells and effervesces in diluted HCl acid

d. contains pebbles and finer sediments

e. has a sooty or shiny appearance

12. The formation of Sample S3 includes:_______.
    a. chemical weathering, transport of ions, precipitation of minerals, lithification
    b. mechanical weathering, transport of sediment a short distance, deposition of sediment, lithification
    c. chemical weathering, transport of ions, precipitation of minerals, lithification, chemical alteration
    d. photosynthesis, growth of organic material, deposition of organic materials, lithifi ation
    e. chemical weathering, transport of ions, precipitation of minerals as shells by organisms, deposition, lithification
13. Sample S4 is called_.
    a. Conglomerate b. Crystalline Limestone
    c. Coal d. Shale
    e. Coquina f. Chert
    g. Dolostone h. Sandstone
14. Sample S4 is an example of a_______.
    a. Clastic Sedimentary Rock b. Organic Sedimentary Rock
    c. Chemical Sedimentary Rock d. Biochemical Sedimentary Rock
15. Closely examine the individual grains in Sample S4. Which of the following is true about its maturity?
    a. It is immature because it is poorly sorted.
    b. It is mature because it is poorly sorted.
    c. It is mature because it contains mostly rounded grains.
    d. It is immature because it contains mostly rounded grains.
    e. It is immature because it has clay-sized particles.
16. The formation of Sample S4 includes:_______.
    a. chemical weathering, transportation of ions, precipitation of minerals, lithification
    b. mechanical weathering, transportation of sediment a short distance, deposition of sediments, lithification
    c. photosynthesis, growth of organic material, deposition of organic materials, lithifi ation
    d. chemical weathering, transportation of ions, precipitation of minerals as shells by organisms, deposition, lithification
17. Sample S5 is called .
    a. Conglomerate b. Crystalline Limestone
    c. Coal d. Shale

e. Coquina f. Chert

g. Dolostone h. Sandstone

18. Sample S5 is an example of a .
    a. Clastic Sedimentary Rock
    b. Organic Sedimentary Rock
    c. Chemical or Biochemical Sedimentary Rock
19. The history of formation of Sample S5 includes: .
    a. chemical weathering, transportation of ions, precipitation of minerals, lithification
    b. mechanical weathering, transportation of sediment a short distance, deposition of sediments, lithification
    c. photosynthesis, growth of organic material, deposition of organic materials, lithifi ation
    d. chemical weathering, transportation of ions, precipitation of minerals as shells by organisms, deposition, lithification
20. Sample S5 has the following characteristic:
    a. effervesces in diluted HCl acid
    b. weakly effervesces in diluted HCl acid if powdered
    c. contains fossil shells and effervesces in diluted HCl acid
    d. contains pebbles and finer sediments
    e. a sooty or shiny appearance
21. Sample S6 is called_.

    a. Conglomerate b. Crystalline Limestone

    c. Coal d. Shale

    e. Coquina f. Chert

    g. Dolostone h. Sandstone
22. The history of formation of Sample S6 includes:_____.
    a. chemical weathering, transportation of ions, precipitation of minerals either biologically or inorganically, lithification
    b. mechanical weathering, transportation of sediments a short distance, deposition of sediments, lithification
    c. photosynthesis, growth of organic material, deposition of organic materials, lithifiation
    d. chemical weathering, transportation of ions, precipitation of minerals as shells by organisms, deposition, lithification

23. Sample S6 is composed of__.
    a. clastic sediments
    b. microcrystalline calcite crystals
    c. microcrystalline dolomite crystals
    d. microcrystalline quartz crystals
    e. organic material
24. Sample S6 can be easily recognized by which of the following properties?
    a. conchoidal fracturing
    b. weakly effervesces in diluted HCl acid if powdered
    c. fissule appearance
    d. a sandpaper texture
    e. a sooty or shiny appearance
25. Sample S7 is called_.
    a. Conglomerate b. Crystalline Limestone
    c. Coal d. Shale
    e. Coquina f. Chert
    g. Dolostone h. Sandstone
26. Compared to Sample S2, how mature is Sample S7?
    a. more mature b. less mature c. same level of maturity
27. Sample S7 consists of______.
    a. fragments of calcite shells b. clay-sized sediments
    c. sand-sized sediments d. organic material
    e. dolomite crystals
28. Sample S7 can be easily recognized by which of the following properties?
    a. conchoidal fracturing
    b. weakly effervesces in diluted HCl acid if powdered
    c. fissile appearance
    d. a sandpaper texture
    e. a sooty or shiny appearance
29. Sample S8 is called_.
    a. Conglomerate b. Crystalline Limestone
    c. Coal d. Shale
    e. Coquina f. Chert
    g. Dolostone h. Sandstone

30. Sample S8 is an example of a_______.

a. Clastic Sedimentary Rock
b. Organic Sedimentary Rock
c. Chemical Sedimentary Rock
d. Biochemical Sedimentary Rock

31. Sample S8 is most likely to weather by which of the following processes?

a. Dissolution
b. Frost Wedging
c. Oxidation
d. Hydrolysis
e. The addition and subtraction of heat

32. The history of formation of Sample S8 includes:_____.

a. chemical weathering, transportation of ions, precipitation of minerals, lithification
b. mechanical weathering, transportation of sediments a long distance, deposition of sediments, lithification
c. mechanical weathering, transportation of sediments a very short distance, deposition of sediments, lithification
d. photosynthesis, growth of organic material, deposition of organic materials, lithification
e. chemical weathering, transportation of ions, precipitation of minerals as shells by organisms, deposition, lithification

33. Search for 39 05 52.46N 84 30 56.16E and zoom out to an eye altitude of ~30,000 ft. The large-scale structures in this sedimentary environment are asymmetrical ripples (known as dunes at this size). Zoom in and out and examine the sedimentary environment. What type of sedimentary environment is this?

a. Lacustrine
b. Fluvial
c. Eolian
d. Glacial

34. What type of weathering do you think is most prominent in this sedimentary environment?

a. Mechanical weathering from a current
b. Dissolution
c. Frost Wedging
d. Hydrolysis

35. Study the large dunes in this image (zoom out to an eye altitude of ~25,000 feet). These structures can indicate the direction that the wind is blowing. What is the predominant wind direction in this area? (Hint: it is easier to see these features if we exaggerate the vertical scale to do this go to Tools, options, and on the 3D view tab change the Elevation Exaggeration to 3. To do this on a Mac go to Google Earth then Preferences)

a. north to south
b. south to north
c. east to west
d. west to east

36. Search for 64 01 03.61N 16 52 56.63W and zoom out to an eye altitude of ~25,000 ft. What type of sedimentary environment is this?

a. Lacustrine b. Fluvial c. Eolian d. Glacial

37. What type of weathering do you think is most prominent in this sedimentary environment? (Make sure you evaluate all choices.)

a. Mechanical weathering from ice b. Dissolution

c. Frost Wedging d. Hydrolysis

e. Both a and c are correct

38. When the sediment you see today lithifies, what type of sedimentary rock would you expect to be most abundant in this area and how mature is this rock type?

a. Shale, Mature b. Shale, Immature

c. Sandstone, Immature d. Sandstone, Mature

e. Conglomerate, Immature f. Conglomerate, Mature

39. Search for 44 42 16.98N 1 05 23.88W and zoom out to an eye altitude of ~17 miles. Notice the depth of this shallow bay (Arcachon Bay, France). This bay dries out at times and later greatly expands in size daily. What sedimentary structure would you likely find in this area because of this?

a. graded beds b. finely layered beds

c. cross beds d. mud cracks

40. What type of sedimentary environment is this?

a. lacustrine b. tidal mudflats

c. deep marine d. beach

41. Search for 20 20 23.94S 150 38 29.14E and zoom out to an eye altitude of ~25 miles (also zoom far out to notice where you are in the world). What type of sedimentary environment is this?

a. shallow marine b. deep marine

c. delta d. reef

42. Think about the origin of this marine formation and consider the latitude to assess climatic conditions. What is the most abundant type of rock you would expect to form in this sedimentary environment?

a. Limestone b. Sandstone c. Coal d. Shale

43. If the sea level dropped 1,000 feet and this sedimentary environment stopped being built and began to break down, what type of weathering would be most likely to occur on these rocks?

a. Fracturing from the addition or subtraction of pressure

b. Dissolution

c. Frost Wedging

d. Hydrolysis

44. Search for 31 00 43.61N 81 25 40.92W and zoom out to an eye altitude of 1,500 ft. What type of sedimentary environment is this?

a. shallow marine
b. reef
c. fluvial
d. each e. tidal mudflat

45. What type of sediment would you expect to find in this depositional environment?

a. Mostly mature clastic sediments with some biochemical and chemical sedimentary sediments

b. Mostly biochemical and chemical sedimentary sediments with some mature clastic sediments

c. Exclusively organic sediments

d. Exclusively biochemical sediments

e. Exclusively immature clastic sediments

11 Chapter
METAMORPHIC ROCKS

11.1 INTRODUCTION: AN OVERVIEW

In any beginning textbook on physical geology, the discourse on metamorphic rocks is positioned subsequent to the chapters on igneous and sedimentary rocks, and for valid reasons. Metamorphic rocks develop by the physical and occasionally chemical transformation of an antecedent rock, be it igneous or sedimentary. In certain instances, metamorphic rocks may be transformed into entirely distinct metamorphic rocks. Igneous rocks originate from the molten material of any rock type, while sedimentary rocks arise from the weathered remnants of any rock type. The transformation of any rock into a metamorphic rock finalizes the elements of the rock cycle. The rocks we currently categorize as igneous, metamorphic, or sedimentary may have previously belonged to a different classification, as rocks undergo recycling over geological time due to tectonic plate movement (refer to Chapter 4). Raising the temperature of a rock can generate magma, and surface rocks can disintegrate into sediment that may eventually lithify into sedimentary rock. How can we transform solid rock into a different type of rock without melting it or converting it into sediment?

All rocks are created under specific temperatures and pressures at or, more typically, beneath the Earth's surface, and they exhibit maximum stability under the conditions of their formation. Consequently, altering the temperature and/or pressure conditions may result in a different rock, one that has adapted to achieve stability under the new external conditions. A metamorphic rock is a type of rock that forms in reaction to alterations in its physical and chemical environment. The term metamorphism denotes a change in form, which for rocks entails the recrystallization of minerals under subsolidus conditions, where temperatures are insufficient for melting. A metamorphic transformation may also transpire if the rock's composition is modified by heated, chemically reactive fluids, resulting in a change in the mineral content of the rock. The word protolith or parent rock refers to the pre-existing rock, distinguishing it from the newly formed metamorphic rock, as all metamorphic rocks possess at least one protolith that has undergone alteration during metamorphism. This chapter elucidates that all metamorphic rocks are classified based on their mineral composition and texture; in metamorphic rocks, texture pertains to the alignment of minerals, while crystal size provides significant insights into the thermal conditions experienced during metamorphism.

In summary, metamorphism is the process through which a pre-existing rock (the protolith) undergoes alteration due to variations in temperature, pressure, or interaction with chemically reactive fluids, or a combination of these factors. The alteration process is a recrystallization event in which the minerals of the original rock have undergone changes in size, shape, and/or composition due to new external conditions. The final outcome is a new metamorphic rock that exhibits a transformed appearance, often markedly distinct from its pre-metamorphic state. The type of metamorphic rock produced is largely contingent upon the original rock prior to the metamorphic process. The agent of metamorphic change is of secondary importance: was recrystallization caused by elevated temperature, pressure, or both, or were chemically reactive fluids involved? Furthermore, it is essential to evaluate the elevated temperatures and pressures, as metamorphism can transpire under various thermal and pressure situations. Metamorphism transpires under subsolidus conditions, indicating that the elevated temperatures facilitating the recrystallization of a rock are insufficient to induce melting.

11.2 AGENTS OF METAMORPHISM

11.2.1 Pressure

All rocks beneath the surface of the earth experience an increase in pressure due to the weight of the overlying sediment and rock layers, and with increasing depth there is a corresponding increase in pressure. This increased pressure does not necessarily cause a rock to become metamorphic, because this particular pressure is typically equal in all directions and is known as lithostatic pressure. Lithostatic pressure is similar to hydrostatic pressure, such as the pressure on the eardrums a swimmer will experience as he or she dives deep in the water. Lithostatic pressure on rocks below the earth's surface may have a change in overall rock volume, but will not cause a change in the shape. An example of decreasing volume due to lithostatic pressure would be a closer packing of clasts and reduction of pore space within a clastic sedimentary rock. But what if the pressure on a rock is un equal, and the rocks become squeezed in one direction more than another direction? This is known as differential pressure, and it can result in a significant change in the appearance of a rock. Figure 11.1 demonstrates how a mineral can change shape due to differential pressure, in this case with the greatest pressures from the top and bottom (as demonstrated by the large gray arrows). Two initially rounded mineral grains (Figure 11.1A) within a sedimentary rock are experiencing the greatest amount of pressure at the contact between the grains (see red arrows in the figure), and the bonds linking the atoms in this grain will break. The atoms will migrate into the area of lesser pressure and reform a bond with other atoms in the mineral grain (Figure 11.1B). As a result, the grains have a flattened shape that is perpendicular to the direction of greatest pressure (Figure 11.1C).

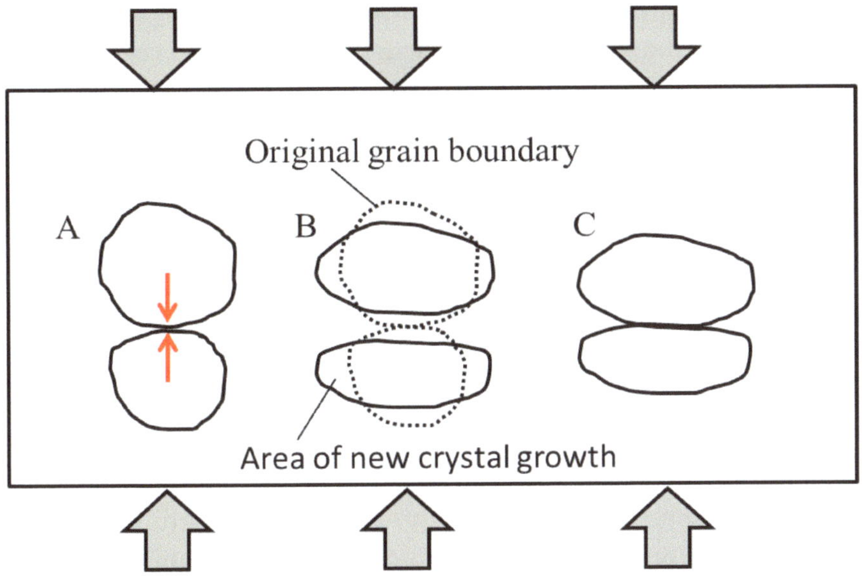

Figure 11.1 | Example of mineral grain deformation due to differential pressure: A) Rounded grains experience the greatest pressure at the contact between the two grains due to the greatest applied pressure as indicated by the large gray arrows. B) The grains change shape by bonds breaking in the high pressure areas, and atoms migrate laterally into the lower pressure areas (the area of new crystal growth). C) The final shape of these two grains. Note that the grains are flattened perpendicular to the direction of greatest pressure.

Figure 11.1 only shows the deformation of two grains; imagine that this is happening to all of the grains in the sedimentary rock, or to all of the phenocrysts (crystals) in an igneous rock. In that case, you will end up with the entire rock having minerals aligned in a certain direction, all by the breaking of bonds between atoms in a mineral, and reforming (recrystallizing) in the lower pressure areas among the grains or crystals in the rock. The end result is a rock with a metamorphic pattern called a **foliation.** Metamorphic foliations are the patterns seen in a rock that has experienced differential pressure; these foliations may be fairly flat, or have a wavy appearance possibly due to more than one direction of greatest pressure. Some rocks may also develop what is called a **lineation**, which can be formed by an elongation of minerals that form a linear feature through the rock. To understand the difference between a foliation and a lineation, let us use some food analogies: a stack of pancakes demonstrates a foliation in your breakfast food, with each pancake layer representing flattened minerals. If you look at the top of the pancakes, you will not see a pattern, but if you view the stack of pancakes from the side, or cut through the stack with your knife in any orientation other than parallel to the pancake layers, you will see the layering or foliation. However, if within that pancake stack, there existed a slice of bacon (yum), the bacon would be the lineation in your breakfast "rock", and you may or may not see it when you cut through the pancake stack. We will discuss foliations, and the different types of foliations, in a later section of this chapter.

11.2.3 Temperature

Probably the most common cause of metamorphism is a change in temperature. Often times metamorphism involves both an increase in temperature along with the pressure changes as described in the above section. Higher temperatures are often associated with metamorphism due to chemically reactive fluids (which we will discuss in the next section). The broad classification for metamorphism into low, medium and high grades of metamorphic change exists mainly due to temperature conditions; this will also be discussed in a later section.

Higher temperatures increase the vibrational energy between the bonds linking atoms in the mineral structure, making it easier for bonds to be broken in order for the recrystallization of the minerals into new crystal shapes and sometimes the development of foliations and lineations as described in the previous section. However, recrystallization can be due to just temperature changes without any differential pressure conditions, and when temperatures are increased, there can be a corresponding increase in mineral sizes as initially small minerals become fused into larger crystals. This fusing of numerous smaller mineral sizes into fewer, yet larger, mineral sizes is known as annealing in metallurgy, but for metamorphic rocks it is still referred to as recrystallization.

In order to understand why increasing temperatures lead to increased grain sizes, we need to again address stability. In general, a mineral grain or crystal is most stable when it has a low surface area to volume ratio, therefore large grains are more stable than small grains, because increasing the grain size results in a greater increase in volume as opposed to a smaller increase in the surface area. Why does stability matter? Because that is why the rocks we are concerned with in this chapter are changing; rocks become unstable when their environment changes, and by a recrystallization processes (metamorphism), they can return to a stable form once again. Figure 11.2 demonstrates the recrystallization process in a sedimentary rock in response to elevated temperature. In this example, the original grains are smaller and rounded, but recrystallization resulted in larger grains that are interlocking; the pore spaces are gone and instead larger crystals exist.

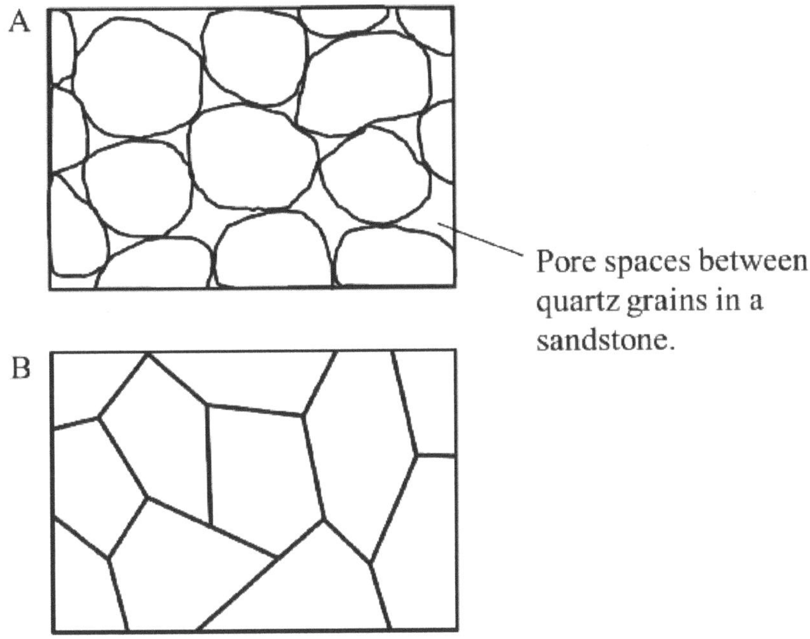

Figure 11.2 | An example of the recrystallization of crystals into larger sizes due to increased temperature: A) a sedimentary rock with rounded quartz grains, and pore spaces between the grains. B) a metamorphic rock with larger, interlocking quartz crystals due to increased temperature conditions.

In addition to increased grain size with increased temperature, occasionally a new mineral forms during metamorphism. These new minerals form at certain temperatures and are called **index minerals**, which can be used to determine the temperature of metamorphism. Index minerals will be covered in more detail in a later section.

11.2.3 Chemically Reactive Fluids

The phrase chemically reactive refers to the dissolved ions in a fluid phase that may react with minerals in a rock; these ions may take the place of some of the atoms in the mineral's structure, which may lead to a significant change in the chemical composition of a rock. Sometimes these fluids are quite hot, especially if they are fluids released from a nearby magma body that is crystallizing while cooling. Metamorphism due to such fluids is known as **hydrothermal metamorphism**.

11.3 METAMORPHIC ROCK NAMES

11.3.1 Foliated Metamorphic Rocks

As mentioned previously, differential pressures can cause a foliation to develop in metamorphosed rocks. There are a few types of foliations that are commonly seen in metamorphic rocks, each foliation type is dependent on the minerals that define the foliation. One type is described as a layering of dark and light colored minerals, so that the foliation is defined as alternating dark and light mineral bands throughout the rock; such a foliation is called **gneissic banding** (Figure 11.3), and the metamorphic rock is called gneiss (pronounced "nice", with a silent g). In Figure 11.3A, the layering in this gneiss is horizontal, and the greatest pressures were at right angles to the gneissic bands. Note that these bands are not always flat, but may be seen contorted as in Figure 11.3B; this rock still is considered to have gneissic banding even though the bands are not horizontal. The typical minerals seen in the dark colored bands are biotite micas and/or amphiboles, whereas the light colored bands are typically quartz or light-colored feldspars. The protoliths for gneiss can be any rock that contains more than one mineral, such as shale with its clay minerals and clay-sized quartz and feldspar, or an igneous rock with both dark-colored ferromagnesian minerals and light-colored non-ferromagnesian minerals (see Chapter 8 for review). In order for gneissic foliation to develop, temperatures

Figure 11.3 | Two examples of the metamorphic rock, gneiss. Each rock exhibits the alternating dark and light mineral bands throughout the rock.

and pressures need to be quite high; for this reason, gneiss rocks represent a high grade of metamorphism. Sometimes a metamorphic rock is seen with mostly amphibole minerals that can define a foliation pattern as well, and is called an amphibolite (Figure 11.4). Occasionally amphibolites also have layers of light-colored plagioclase minerals present; in that case, the rock may be called an amphibole gneiss. Amphibolites may not always have a foliation, however the rock picture in Figure 11.4 appears to have a parallel alignment of the amphiboles (see red arrows in figure), making this a foliated amphibolite. The protolith for an amphibolite must be a rock type with a large amount of dark ferromagnesian minerals, such as a mafic or ultramafic igneous rock. Amphibolites require higher temperatures and pressures to form as well, and are considered to be high grade metamorphic rocks.

Another type of foliation is defined by the presence of fl or platy minerals, such as muscovite or biotite micas. Metamorphic rocks with a foliation pattern defined by the layering of platy minerals are called schist; the rock name is commonly modified to indicate what mica is present. For example, Figure 11.5 is a photo of a muscovite schist, however it also has garnet present, so the correct name for the rock pictured in Figure 11.5 is a garnet muscovite schist. By convention, when naming a metamorphic rock the mineral in the lowest quantity (garnet, in this case) is mentioned first. Notice that the muscovite micas

Figure 11.4 | Example of a foliated amphibolite. Notice the alignment of the amphibole minerals, best seen where the red arrows are indicating. The lighter colored minerals in this rock are plagioclase feldspars.

Figure 11.5 | Photo of a garnet muscovite schist. A) The muscovite micas are large enough to be seen as very shiny minerals in the top photograph. B) Side view of the schist shows the distinctive wavy foliation made by the platy micas. This metamorphic rock also contains the index mineral, garnet.

define a very wavy foliation in the rock; this textural pattern of wavy micas is called a **schistose foliation** (Figure 11.5B). The sedimentary rock shale is usually the protolith for schist; during metamorphism, the very tiny clay minerals in shale recrystallize into micas that are large enough to see unaided. Temperatures and pressures necessary for schistose foliation are not as high as those for gneiss and amphibolite; therefore schists represent an intermediate grade of metamorphism.

There is one foliation type that is defined by the alignment of minerals that are too small to see, yet the foliation can still be visible. This type of foliation is only seen in the metamorphic rock called slate; slate forms by the low temperature and low pressure alteration of a shale protolith. The clay sized minerals in the shale recrystallize into very tiny micas which are larger than the clay minerals, but still too small to be visible. However, because these tiny micas are aligned, they control how the metamorphic rock (slate) breaks, and the rock tends to break parallel to the mica alignment. Therefore, even though we cannot see the aligned minerals that define the foliation, we can use the alignment of the rock fracture pattern, as the rock is cleaved or split. For this reason, the foliation is called a **slaty cleavage**, and a rock displaying this type of foliation is called a slate. Figure 11.6 is an example of the foliated slate displaying slaty cleavage; notice that this rock has retained its original sedimentary layering (depositional beds), which in this case is quite different from the foliation direction. The only protolith for slate is shale, and the fact that original sedimentary features and even some fossils in shale may be preserved and visible in slate is due to the low temperatures and pressures that barely alter the shale protolith, making slate an example of a low grade metamorphic rock. Slate has great economic value in the construction industry; due to its ability to break into thin layers and impermeability to water, slate is used as roofing tiles and flooring.

Figure 11.6 | Slaty cleavage in the metamorphic rock, slate. The alignment of very tiny micas controls how this rock breaks or cleaves. Note the banded appearance on the top of this rock, which is the depositional layering of the original rock, shale.

11.3.2 Non-Foliated Metamorphic Rocks

The other class of metamorphic rocks is non-foliated; the lack of foliation may be due to a lack of differential pressure involved in the metamorphic process, however this is not necessarily the case for all non-foliated metamorphic rocks. If the protolith rock is monomineralic (composed of one mineral type), such as limestone, dolostone, or a sandstone with only quartz sands, then a foliation will not develop even with differential pressure. Why? The calcite mineral in limestone, the dolomite mineral in dolostone, and the quartz sands in sandstone are neither platy minerals, nor are there different colored minerals in these rocks. These minerals (calcite, dolomite, and quartz) recrystallize into equigranular, coarse crystals (see Figure 11.2B), and the metamorphic rocks that they make are named by their composition, not by foliation type. For example, Figure 11.7 is quartzite, a metamorphosed quartz-rich sandstone. Figure 11.8 shows two examples of marble; note that color can vary for marble, as well as for the quartzite. As a result, quartzites and marbles may be hard to identify based on appearance, therefore you must rely on the properties of the minerals that comprise these rocks; you may recall that quartz is harder than glass, while limestone and dolomite are softer than glass. Also, marble will react (effervesce) to acid, but quartz will not react. If you zoom in for a close view of the marble in Figure 11.8, you will see the calcite crystals are fairly large compared to the quartz crystals in the quartzite in Figure 11.7; this can be attributed to the temperature of metamorphism, as higher temperatures result in larger crystals. These rocks are also of economic importance; marble and quartzite are used for dimension stone in buildings and for countertops in many homes. Furthermore, marble is commonly used for statues and sometimes grave markers.

Figure 11.7 | An example of quartzite; a nonfoliated metamorphic rock. Quartzite can appear in a variety of colors, but most are fairly light in color.

Figure 11.8 | Examples of the non-foliated rock, marble. The pink and white colors are common, however some marbles are darker in color, and others may be white with dark gray markings (like the famous Carrara marble from Italy).

One final non-foliated rock type that should be mentioned is anthracite coal (Figure 11.9). As you may recall, coal is a sedimentary rock composed of fossilized plant remains. This sedimentary coal is called bituminous coal; under higher temperatures and pressures bituminous coal can lose more of the volatiles typical of coal (water vapor, for example), but the carbon content is enriched, making metamorphic coal (anthracite coal) a hotter burning coal due to the higher carbon content. Anthracite coal can be distinguished from sedimentary coal by the shinier appearance, and is somewhat harder than bituminous coal, although both coal types are of low density due to their carbon content. Note that this particular metamorphism is not a recrystallization event, per se, as coal is mostly organic remains. However, bonds are still being broken and reforming due to changes in temperature and pressure.

In order to identify and name metamorphic rocks, a logical first step would be to examine the rock for evidence of any pattern or foliation, and if present, identify what mineral or minerals are making the foliation pattern. Non-foliated metamorphic rocks can be identified by the properties defined by their mineral composition. Below is a table summarizing the metamorphic rock types, foliation names, and protolith rock types (Table 11.1).

Figure 11.9 | Anthracite coal. Note the shiny appearance of this metamorphosed coal.

Table 11.1 | Summary chart of metamorphic rocks discussed in this chapter, including the names of some of the possible sedimentary rock and igneous rock protoliths for each metamorphic rock. A metamorphic rock can also be a protolith: for example, a slate can be a protolith for a schist.

Texture	Characteristics	Protolith	Metamorphic Rock Name	
Foliated (Increasing Temperature & Pressure ↓)	Fine-grained rock, tends to split in parallel fragments (known as slaty cleavage)	Shale	**Slate**	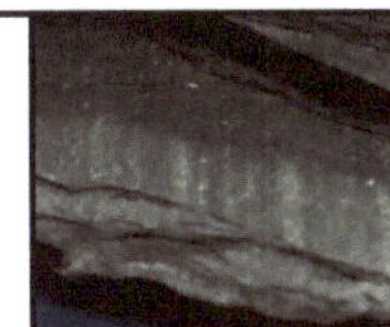
	Contains shiny muscovite (light) or biotite (dark) micas, may see other minerals. Has schistose pattern of foliation	Shale	**Schist**	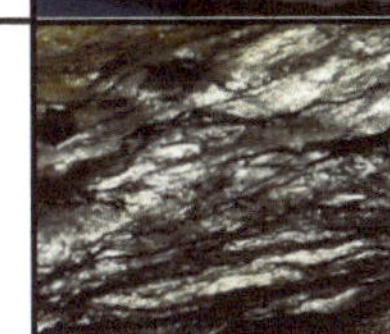
	Contains alternating bands of light- and dark-colored minerals (usually biotite or amphibole), called gneissic banding.	Shale or Igneous Rock	**Gneiss**	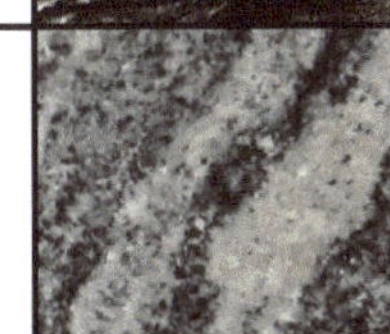
Possibly Foliated	Most of the minerals in this rock are amphiboles, which may be aligned to form a foliation.	Mafic or Ultramafic Rock	**Amphibolite**	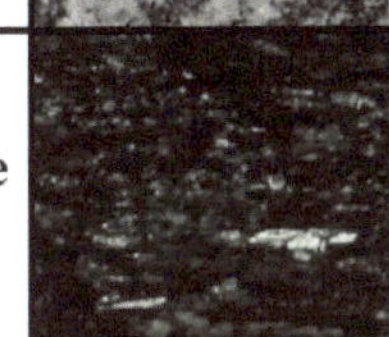
Non-foliated	Equigranular grains of quartz, which has a hardness of 7	Sandstone or Siltstone	**Quartzite**	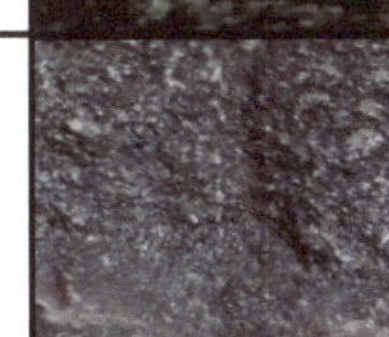
	Equigranular grains of calcite or dolomite, which has a hardness of 4, and reacts to acid	Limestone or Dolostone	**Marble**	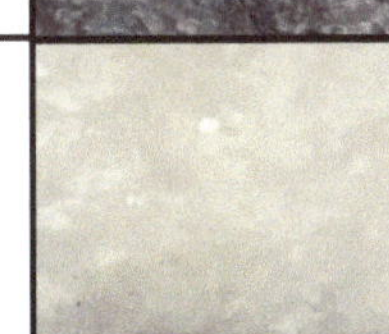
	Contains mostly carbon, is a low density rock, and has a shiny appearance.	Bituminous Coal	**Anthracite Coal**	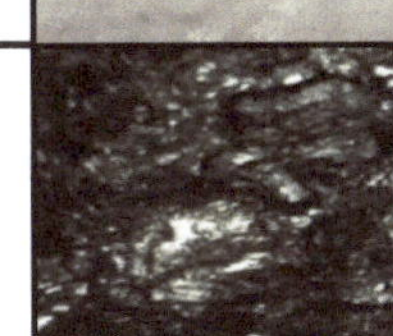

11.3.3 LAB EXERCISE

Part A Metamorphic Rock Identification

In order to answer the questions in this lab, you will need the rock samples from your HOL rock kit that are labeled M1 through M7 (located in the metamorphic rock bag). A photo of these samples is given in Figure 11.10. You will also need your hand lens, glass plate, and the HCl bottle.

Figure 11.10 | Metamorphic rocks M1 through M7.

1. Sample M1 has the following texture:
   a. slaty cleavage b. schistose foliation
   c. gneissic banding d. lineation
   e. non-foliated
2. Sample M1 is called:
   a. marble b. quartzite c. schist d. gneiss
   e. slate f. anthracite coal g. amphibolite
3. The dark minerals in Sample M1 are:
   A. flat biotite micas b. tabular amphiboles
   c. flat muscovite micas d. tabular feldspars
4. A possible protolith for Sample M1 is:
   a. an ultramafic rock b. basalt
   b.c. limestone d. granite
   c.e. sandstone
5. Sample M2 has the following texture:
   a. slaty cleavage b. schistose foliation
   c. gneissic banding d. lineation
   e. non-foliated
6. Sample M2 is called:
   a. marble b. quartzite c. schist d. gneiss
   e. slate f. anthracite coal g. amphibolite

7. Sample M3 has the following texture:
   a. slaty cleavage b. schistose foliation
   c. gneissic banding d. lineation
   e. non-foliated

8. Sample M3 is called:
   a. marble b. quartzite c. schist d. gneiss
   e. slate f. anthracite coal g. amphibolite

9. A possible protolith for Sample M3 is:
   a. bituminous coal b. basalt c. shale
   d. clay e. rhyolite f. sandstone

10. Sample M3 is an example of:
    a. high grade of metamorphism
    b. intermediate grade of metamorphism
    c. low grade of metamorphism

11. Sample M3 is used for:
    a. road construction b. statues c. roof tiles d. barbequing

12. Sample M4 has the following texture:
    a. slaty cleavage b. schistose foliation
    c. gneissic banding d. lineation
    e. non-foliated

13. Sample M4 is called:
    a. marble b. quartzite c. schist d. gneiss
    e. slate f. anthracite g. amphibolite

14. Sample M4 is mainly composed of this mineral:
    a. biotite mica b. amphibole c. muscovite mica
    d. calcite e. quartz

15. Sample M4 is an example of:
    a. high grade of metamorphism
    b. intermediate grade of metamorphism
    c. low grade of metamorphism

16. Sample M5 has the following texture:
    a. slaty cleavage b. schistose foliation
    c. gneissic banding d. lineation
    e. non-foliated

17. Sample M5 is called:
    a. marble b. quartzite c. schist d. gneiss
    e. slate f. anthracite g. amphibolite
18. Sample M5 is mainly composed of this mineral:
    a. biotite mica b. amphibole c. muscovite mica
    d. calcite e. quartz
19. A possible protolith for Sample M5 is:
    a. diorite b. quartz c. limestone d. shale e. sandstone
20. Sample M6 is mainly composed of this dark mineral:
    a. biotite mica b. amphibole c. muscovite mica
    d. calcite e. quartz
21. Sample M6 is called:
    a. marble b. quartzite c. schist d. gneiss
    e. slate f. anthracite g. amphibolite
22. A possible protolith for Sample M6 is:
    a. basalt b. granite c. sandstone
    d. limestone e. shale
23. Sample M6 is an example of:
    a. high grade of metamorphism
    b. intermediate grade of metamorphism
    c. low grade of metamorphism
24. Sample M7 has the following texture:
    a. slaty cleavage b. schistose foliation c. gneissic banding
    d. lineation e. non-foliated
25. Sample M7 is called:
    a. marble b. quartzite c. schist d. gneiss
    e. slate f. anthracite g. amphibolite
26. A possible protolith for Sample M7 is:
    a. basalt b. granite
    c. sandstone d. limestone
    e. shale

11.4 TYPES OF METAMORPHISM

11.4.1 Regional Metamorphism

Metamorphism that influences entire rock masses across an extensive area is termed regional metamorphism. A diverse array of temperature and pressure circumstances yields several metamorphic rock types; for instance, all foliated rocks belong to this metamorphic category, along with certain non-foliated rocks. A significant region is influenced by variations in temperature and/or pressure. A specific form of regional metamorphism transpires along convergent plate borders, where rocks experience varying pressures and temperatures, leading to thermodynamic metamorphism—"thermo-" denoting elevated temperature and "dynamic" indicating heightened pressure. Mountain ranges formed by the convergence of tectonic plates exemplify thermodynamic metamorphism, such as the Himalayan Mountains in Asia and the Sierra Nevada Mountains in North America. The mountain range names can be accessed by entering them in the Google Earth search field. Regional metamorphism transpires at plate boundaries where an oceanic plate subducts into the mantle due to plate convergence, as previously discussed in the plate tectonics chapter. The subduction of oceanic plates results in the formation of deep ocean trenches, exemplified by the trench along the western margin of South America. In locations of heightened pressure, metamorphic foliations can become distinctly prominent, as seen by the progression from slate to schist to gneiss; concurrently, elevated temperatures result in larger grain sizes and the emergence of new minerals within the metamorphosed rock. These novel minerals are significant as they delineate the metamorphic temperature for both regional and contact metamorphism.

Figure 11.11 exemplifies the potential index minerals that develop during the metamorphism of a shale protolith. Shale is a sedimentary rock composed of clay minerals, clay-sized quartz, and occasionally feldspars. As temperature rises, recrystallization leads to the emergence of new minerals, first with chlorite at the lowest metamorphic temperatures. High-temperature metamorphism leads to the creation of the mineral sillimanite. The metamorphic portions depicted in this picture are referred to as mineral zones, designated according to the appearance of each distinct mineral. The garnet zone commences with the initial presence of garnet in the rock, while the emergence of the mineral staurolite signifies the conclusion of the garnet zone and the onset of the staurolite zone. The examination of index mineral zones is significant for economic exploration, as certain index minerals possess economic value; for instance, garnet is utilized as an abrasive and occasionally as a decorative gemstone.

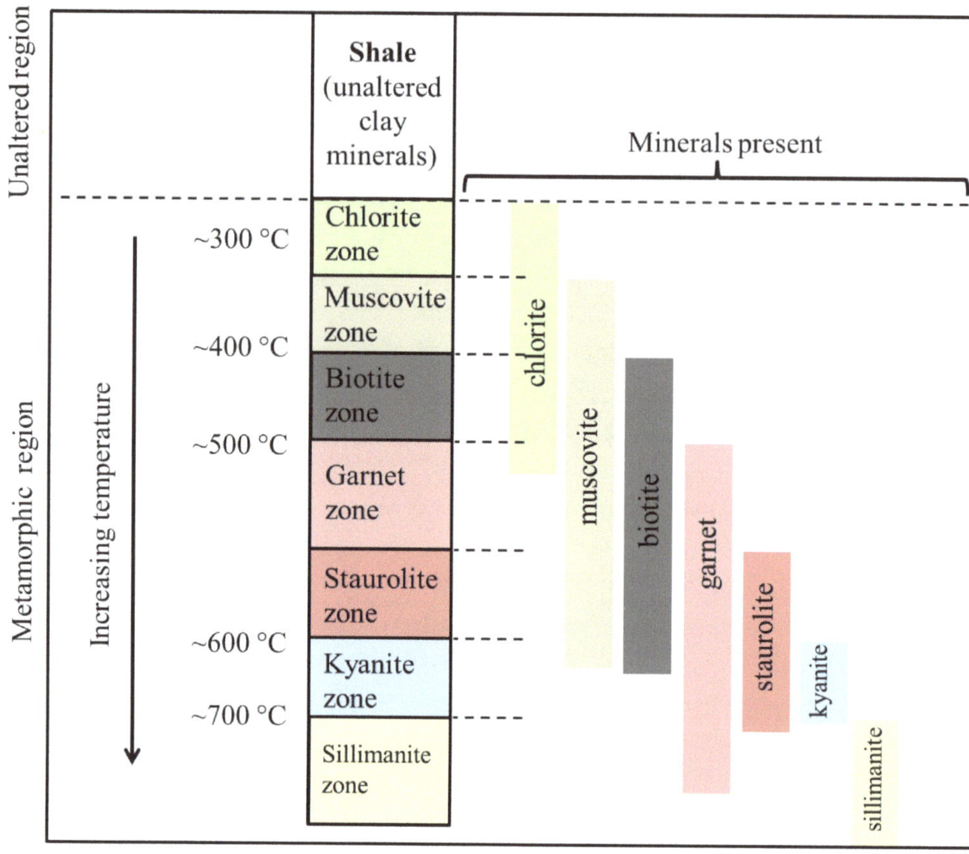

Figure 11.11 | Index mineral chart, showing possible minerals that can form from a shale protolith. The appearance of certain index minerals is also pressure dependent.

11.4.2 Contact Metamorphism

As mentioned previously, recrystallization due to increased temperature results in the formation of larger minerals, or sometimes in the formation of new minerals (Figure 11.11). **Contact metamorphism** describes the type of metamorphism attributed to increased temperature, usually from proximity to a heat source such as an intrusive magma body or a lava flow, and the rock undergoes a **thermal metamorphism**. The rocks that are in closer contact to the magma will form larger crystals due to the higher heat, and may form high temperature index minerals such as garnet, staurolite, kyanite and sillimanite. Rocks that are in contact with hot chemically reactive fluids can also fall within the contact metamorphism category; recrystallization due to this type of contact is called **hydrothermal metamorphism**. Unlike the regional metamorphism described in the previous section, both types of contact metamorphism are confided to smaller areas near the heat source (magma, lava, or hydrothermal fluids). Since differential pressures are not involved, contact metamorphism results in the formation of non-foliated rocks. The non-foliated rock types that were covered previously included marble, quartzite, amphibolite, and anthracite coal. Other non-foliated rocks that have undergone contact metamorphism are broadly classified as hornfels or granofels, to indicate that they are recrystallized rocks that lack foliation; for example, shale can be altered into a hornfels if it is recrystallized, but lacks slaty cleavage due to the lack of any developed foliation.

11.4.3 Shock (Impact) Metamorphism

This type of metamorphism is not as common as regional or contact metamorphism, and it differs from the other types in a spectacular way. Regional and contact metamorphism requires quite a bit of time (up to several hundred thousand years) for rocks to metamorphose; **shock metamorphism** takes place in a matter of seconds. This type of metamorphism occurs at the site of meteorite craters; rocks at the surface and near surface are subjected to extreme stresses upon impact, and sometimes portions of these surface rocks become tiny blobs of melt that cool to form a type of glass called a **tektite**. Age dating of these tektites has helped in the determination and even recognition of ancient meteorite impact sites worldwide. Shock or impact metamorphism is mainly due to the extreme stresses brought on by the impact, and for this reason shock metamorphism is considered to be a **dynamic metamorphism**, as opposed to dynamothermal (regional) metamorphism.

11.4.4 Fault Zone Metamorphism

Rocks along geologic faults are deformed due to pressures associated with shearing, compression, or extensional stresses, with minor changes due to heat.

Figure 11.12 | Example of a mylonite rock; note the deformed lightcolored minerals (feldspars) that have developed a tail (red arrow). The direction of shearing can sometimes be determined by examining these minerals.

As with shock metamorphism, **fault zone metamorphism** is a type of dynamic metamorphism. Shallow faults (close to the earth's surface) may grind the nearby rocks into smaller, angular fragments called fault breccia, or you may see a fine clay powder called fault gouge; fault gouge forms by the chemical alteration of fault breccia. Fault breccia is a form of brittle deformation of rock; at greater depths, rocks along a fault will undergo a more plastic type of deformation. The rocks that form adjacent to the fault in deeper sections are called **mylonite** (Figure 11.12). The minerals in mylonitic rocks are deformed due to shear stresses associated with movement along the fault.

11.4.5 LAB EXERCISE

Part B Types of Metamorphism and Metamorphic Regions

For this portion of the lab, you will be using Google Earth, and Figure 11.11 in this chapter. You will also need your rock samples M1 and M7.

27. According to Figure 11.11, what mineral is unstable in the staurolite zone?

    a. chlorite b. muscovite c. biotite d. garnet

28. Refer again to Figure 11.11; which index minerals will not be found together in the same rock?

    a. chlorite and garnet b. chlorite and kyanite

    c. biotite and staurolite d. biotite and muscovite

29. Sample M1 from your rock kit is most likely an example of:

    a. hydrothermal metamorphism b. contact metamorphism

    c. regional metamorphism d. shock metamorphism

30. Sample M7 from your rock kit is most likely an example of:

    a. fault zone metamorphism b. contact metamorphism

    c. regional metamorphism d. shock metamorphism

31. Type 8 56 22.28N 126 55 54.33E in the search bar of Google Earth. Zoom out to an eye altitude of ~1700 miles. What type of metamorphism is likely in this area?

    a. shock metamorphism b. contact metamorphism

    c. regional metamorphism d. dynamic metamorphism

32. The agents of metamorphism in this region, are most likely:

    a. chemically reactive fluids b. temperature

    c. pressure d. both temperature and pressure

33. Type 35 06 53.42N 119 38 47.59W in the search bar of Google Earth. Zoom out to an eye altitude of ~8500ft. What type of metamorphism is likely in this area?

    a. fault zone metamorphism b. contact metamorphism

    c. regional metamorphism d. shock metamorphism

34. Evidence of metamorphism at this location could be from the discovery of:

a. gneiss b. quartzite c. tektites d. mylonite

35. Type 35 01 38.93N 111 01 21.93W in the search bar of Google Earth. Zoom out to an eye altitude of ~12,000ft. What type of metamorphism is likely in this area?

a. fault zone metamorphism b. contact metamorphism

c. regional metamorphism d. shock metamorphism

36. Evidence of metamorphism at this location could be from the discovery of:

a. gneiss b. quartzite c. tektites d. mylonite

11.5 STUDENT RESPONSES

1. Sample M1 has the following texture:

a. slaty cleavage b. schistose foliation c. gneissic banding

d. lineation e. non-foliated

2. Sample M1 is called:

a. marble b. quartzite c. schist d. gneiss

e. slate f. anthracite g. amphibolite

3. The dark minerals in Sample M1 are:

a. flat biotite micas b. tabular amphiboles

c. flat muscovite micas d. tabular feldspars

4. A possible protolith for Sample M1 is:

a. an ultramafic rock b. basalt

c. limestone d. granite

e. sandstone

5. Sample M2 has the following texture:

a. slaty cleavage b. schistose foliation c. gneissic banding

d. lineation e. non-foliated

6. Sample M2 is called:

a. marble b. quartzite

c. schist d. gneiss

e. slate f. anthracite coal

g. amphibolite

7. Sample M3 has the following texture:

a. slaty cleavage b. schistose foliation

c. gneissic banding d. lineation

e. non-foliated

8. Sample M3 is called:
   a. marble b. quartzite c. schist d. gneiss
   e. slate f. anthracite g. amphibolite
9. A possible protolith for Sample M3 is:
   a. bituminous coal b. basalt
   c. shale d. clay
   e. rhyolite f. sandstone
10. Sample M3 is an example of:
    a. high grade of metamorphism
    b. intermediate grade of metamorphism
    c. low grade of metamorphism
11. Sample M3 is used for:
    a. road construction b. statues
    c. roof tiles d. barbequing
12. Sample M4 has the following texture:
    a. slaty cleavage b. schistose foliation
    c. gneissic banding d. lineation
    e. non-foliated
13. Sample M4 is called:
    a. marble b. quartzite c. schist d. gneiss
    e. slate f. anthracite g. amphibolite
14. Sample M4 is mainly composed of this mineral:
    a. biotite mica b. amphibole c. muscovite mica
    d. calcite e. quartz
15. Sample M4 is an example of:
    a. high grade of metamorphism
    b. intermediate grade of metamorphism
    c. low grade of metamorphism
16. Sample M5 has the following texture:
    a. slaty cleavage b. schistose foliation
    c. gneissic banding d. lineation
    e. non-foliated

17. Sample M5 is called:
    a. marble b. quartzite c. schist d. gneiss
    e. slate f. anthracite g. amphibolite
18. Sample M5 is mainly composed of this mineral:
    a. biotite mica b. amphibole c. muscovite mica
    d. calcite e. quartz
19. A possible protolith for Sample M5 is:
    a. diorite b. quartz c. limestone d. shale e. sandstone
20. Sample M6 is mainly composed of this dark mineral:
    a. biotite mica b. amphibole
    c. muscovite mica d. calcite
    e. quartz
21. Sample M6 is called:
    a. marble b. quartzite c. schist d. gneiss
    e. slate f. anthracite g. amphibolite
22. A possible protolith for Sample M6 is:
    basalt b. granite c. sandstone d. limestone e. shale
23. Sample M6 is an example of:
    a. high grade of metamorphism
    b. intermediate grade of metamorphism
    c. low grade of metamorphism
24. Sample M7 has the following texture:
    a. slaty cleavage b. schistose foliation
    c. gneissic banding d. lineation
    e. non-foliated
25. Sample M7 is called:
    a. marble b. quartzite c. schist d. gneiss
    e. slate f. anthracite g. amphibolite
26. A possible protolith for Sample M7 is:
    a. basalt b. granite
    c. sandstone d. limestone e. shale
27. According to Figure 11.11, what mineral is unstable in the staurolite zone?
    a. chlorite b. muscovite c. biotite d. garnet

28. Refer again to Figure 11.11; which index minerals will not be found together in the same rock?
    a. chlorite and garnet
    b. chlorite and kyanite
    c. biotite and staurolite
    d. biotite and muscovite
29. Sample M1 from your rock kit is most likely an example of:
    a. hydrothermal metamorphism
    b. contact metamorphism
    c. regional metamorphism
    d. shock metamorphism
30. Sample M7 from your rock kit is most likely an example of:
    a. fault zone metamorphism
    b. contact metamorphism
    c. regional metamorphism
    d. shock metamorphism
31. Type 8 56 22.28N 126 55 54.33E in the search bar of Google Earth. Zoom out to an eye altitude of ~1700 miles. What type of metamorphism is likely in this area?
    a. shock metamorphism
    b. contact metamorphism
    c. regional metamorphism
    d. dynamic metamorphism
32. The agents of metamorphism in this region, are most likely:
    a. chemically reactive fluids
    b. temperature
    c. pressure
    d. both temperature and pressure
33. Type 35 06 53.42N 119 38 47.59W in the search bar of Google Earth. Zoom out to an eye altitude of ~8500ft. What type of metamorphism is likely in this area?
    a. fault zone metamorphism
    b. contact metamorphism
    c. regional metamorphism
    d. shock metamorphism
34. Evidence of metamorphism at this location could be from the discovery of:
    a. gneiss
    b. quartzite
    c. tektites
    d. mylonite
35. Type 35 01 38.93N 111 01 21.93W in the search bar of Google Earth. Zoom out to an eye altitude of ~12,000ft. What type of metamorphism is likely in this area?
    a. fault zone metamorphism
    b. contact metamorphism
    c. regional metamorphism
    d. shock metamorphism
36. Evidence of metamorphism at this location could be from the discovery of:
    a. gneiss
    b. quartzite
    c. tektites
    d. mylonite

12 Chapter

CRUSTAL DEFORMATION

12.1 INTRODUCTION: AN OVERVIEW

The Earth is a dynamic globe influenced by active forces. Such forces can create mountains and deform rocks by crumpling and folding. Rocks undergo deformation in response to these stresses, resulting in alterations in their shape and/or volume. The resultant characteristics are referred to as geologic structures. This deformation can create striking and aesthetically pleasing landscapes, as illustrated in Figure 12.1, which depicts the alteration of previously horizontal rock strata.

What is the significance of studying crustal deformation? Such investigations can furnish us with a chronicle of the past and the factors that were at play during that time. The accurate interpretation of features formed during deformation is essential in the petroleum and mining sectors. It is also crucial for engineering. Comprehending the behavior of deformed rocks is essential for the design and preservation of secure engineered constructions.

Figure 12.1 | Rocks that have been deformed along the coast of Italy

When proper geological planning is not considered in engineering, disasters can strike. For example, the Vajont Dam was constructed at Monte Toc, Italy in the early 1960's. The place was a poor choice for a dam, as the valley was narrow, thorough geological tests were not performed, and the area surrounding the dam was prone to large landslides. The steep canyon walls were composed of limestone with solution cavities, not known for its stability, and shifting and fracturing of rock that occurred during the filling of the reservoir went unheeded. In 1963, a massive landslide in the area displaced much of the water in the dam, causing it to override the top of the dam and flood the many villages downstream, resulting in the deaths of almost 2,000 people (Figure 12.2).

12.1.1 Learning Outcomes

Figure 12.2 | An image of the Vajont reservoir shortly after the massive landslide (you can see the scar from the landslide on the right, and the dam is located in the foreground on the left).

After completing this chapter, you should be able to:

- Understand the different types of stress that rocks undergo, and their responses to stress
- Demonstrate an understanding of the concepts of strike and dip
- Recognize the different types of folds and faults, and the forces that create them
- Use block diagrams to display geologic features
- Create a geologic cross-section

12.2 STRESS AND STRAIN

Rocks change as they undergo **stress**, which is just a force applied to a given area. Since stress is a function of area, changing the area to which stress is applied makes a difference. For example, imagine the stress that is created both at the tip of high heeled shoes and the bottom of athletic shoes. In the high heeled shoe, the area is very small, so that stress is concentrated at that point, while the stress is more spread out in an athletic shoe. Rocks are better able to handle stress that is not concentrated in one point. There are three main types of stress: compression, tension, and shear. When **compressional forces** are at work, rocks are pushed together. **Tensional forces** operate when rocks pull away from each other. **Shear forces** are created when rocks move horizontally past each other in opposite directions. Rocks can withstand compressional stress more than tensional stress (see Figure 12.3).

Applying stress creates a deformation of the rock, also known as strain. As rocks are subjected to increased stress and strain, they at first behave in an elastic manner, which means they return to their original shape after deformation (Figure 12.4). This elastic behavior continues until rocks reach their elastic limit (point X on Figure 12.4), at which point plastic deformation commences. The rocks may bend into folds, or behave in a brittle manner by fracturing (brittle behavior can be easily envisioned if you think of a hammer hitting glass), but regardless they do not return to their original shape when the stress is removed in plastic deformation. The resulting deformation from applied stress depends on many factors, including the type of stress, the type of rock, the depth of the rock and pressure and temperature conditions, and the length of time the rock endures the stress. Rocks behave very differently at depth than at the surface. Rocks tend to deform in a more plastic manner at depth, and in a more brittle manner near the Earth's surface.

Figure 12.3 | This is a picture of the Roman Forum. Why did the Romans use so many vertical columns to hold up the one horizontal beam? If the horizontal beam spanned a long distance without support, it would buckle under its own weight. This beam is under tensional stress, so it is not as strong.

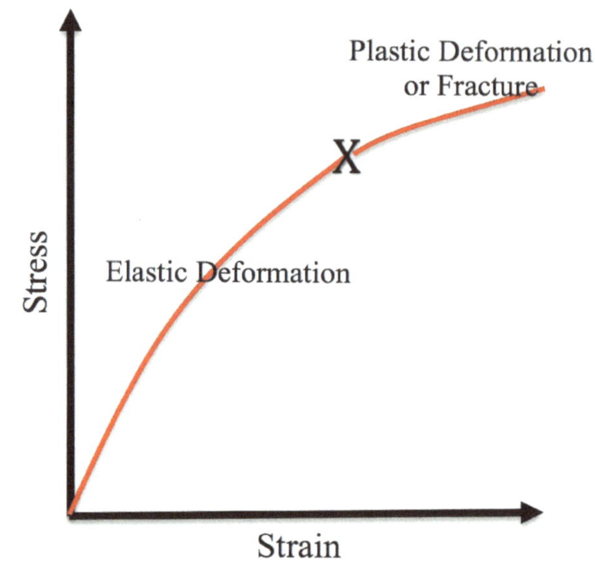

Figure 12.4 | A stress and strain diagram. As stress and strain increase, rocks fi experience elastic deformation (and can return to their original shape) until the elastic limit is reached, depicted at point X. After this point, rocks will fracture or experience plastic deformation, so that their original shape is destroyed

12.3 STRIKE AND DIP

To learn many of the concepts associated with structural geology, it is useful to use block diagrams. As you examine these blocks, note the different ways that you can view them. If you look at a block from along the side, you are seeing the cross-section view (like what you see along roads that have been cut through the mountains). If you look at the block from directly above it, you are looking at map view (Figure 12.5).

As you think about how rocks have changed through the process of deformation, it is useful to remember how they deposited in the first place. We will briefly review some of the geologic laws that you learned in the Introduction Lab. Sedimentary rocks, under the influence of gravity, will deposit in horizontal layers (Law of Original Horizontality). The oldest rocks will be on the bottom (because they had to be there first for the others to deposit on top of them), and are numbered with the oldest being #1 (Law of Superposition). The wooden block in Figure 12.6 displays how depends on many factors, including the type of stress, the type of rock, the depth of the rock and pressure and temperature conditions, and the length of time the rock endures the stress. Rocks behave very differently at depth than at the surface. Rocks tend to deform in a more plastic manner at depth, and in a more brittle manner near the Earth's surface. Sedimentary rocks originally deposit. Each of the boundaries between the colored rock units represents a geologic **contact**, which is simply the surface between two different rock units. Earth's rock layers are usually not this uncomplicated. Rock layers are often at an angle, not horizontal, indicating that changes have occurred since deposition. Examine Figure 12.7 to see tilted rocks.

Figure 12.5 | You are viewing the top block in this image in map view, viewed from directly above the block. The lower block is from the same rock layers, and you are viewing it in cross-section (or from the side). Note that in cross-section, you can see how the rock layers are tilted.

Figure 12.6 | In this image, different rock types are given different colors. The oldest rock, on the bottom, is given a #1. The youngest rock in this image is #4. Author: Randa Harris

Figure 12.7 | In this image, beds have been tilted. Which color bed is the oldest? To determine this, it is useful to apply Occam's Razor, which states that the simplest explanation is most likely the best. It is more likely that the gray bed on the left was the bottom bed

In order to measure and describe layers like this, geologists apply the concepts of strike and dip. **Strike** refers to the line formed by the intersection of a horizontal plane and an inclined surface. **Dip** is the angle between that horizontal plane (such as the top of this block) and the tilted surface (the geologic contact between the tilted layers). In Figure 12.8, look at the tilted sedimentary layers. Strike is a line on the horizontal plane created when the dipping green layer intersects the Earth's surface. To better display the horizontal surface, it has been represented by water. The dip angle is measured from the horizontal surface to the dipping bed (outlined in orange on this image).

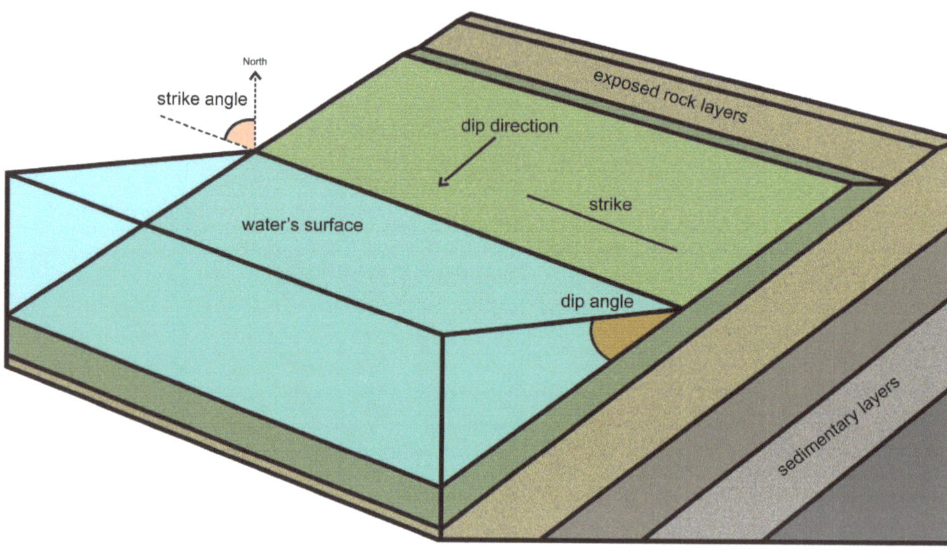

Figure 12.8 | A demonstration of strike and dip. In this example, the beds are dipping to the southwest.

Now, let's apply this concept to the block of dipping beds that you just looked at (Figure 12.7). To determine strike, find where the dipping layer intersects the horizontal surface and draw a line parallel to this line of intersection on the top of the block (i.e. our horizontal surface). To determine dip, pretend that there is a drop of water between one bed and the next, for example, along the intersection of the pale blue bed and the red bed. In which direction would the water roll if it followed that contact? That is the direction of dip—towards the right in this case. The symbol for strike and dip is given along the top of the block. Note that the dip symbol (shorter line) is perpendicular to the strike symbol (Figure 12.9).

Useful in map interpretation is the application of the Rule of V's to determine dip direction. As a stream crosses tilted strata, it will cut a V shape into the rock layers. The point of the V is in the direction of dip. In vertical strata, no V shapes are created (Figure 12.10).

Figure 12.9 | This image depicts the strike and dip for the pale blue bed. As all the beds are oriented in the same direction, they would all have the same strike and dip.

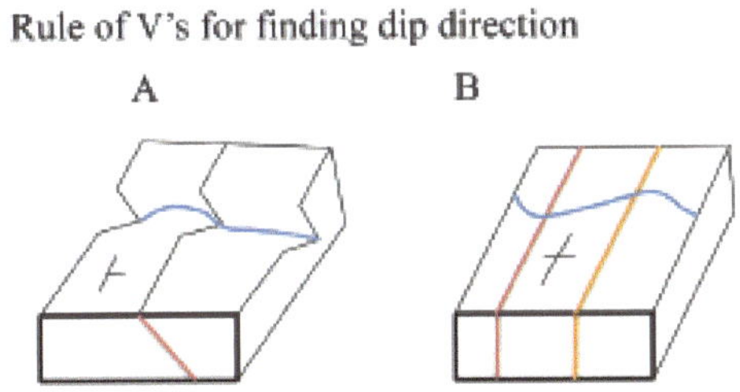

Figure 12.10 | In image A, the stream is encountering tilted strata, which results in a V shape that points towards the dip direction (in this case, to the right). In image B, the strata are vertical, so the stream cannot create a V shape. For reference on common geologic map symbols, refer to Figure 12.31 at the end of this lesson.

12.3.1 LAB EXERCISE

Part A (6 points)

1. For the following diagram, determine the correct map symbol that would go in the oval box.

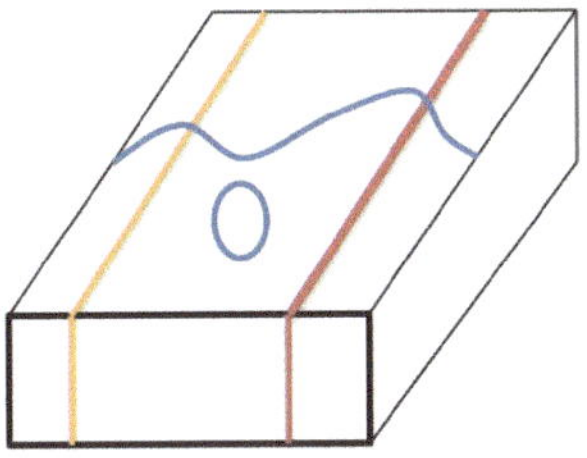

a.  b.  c.  d. 

2. For the following diagram, determine the correct map symbol that would go in the oval box.

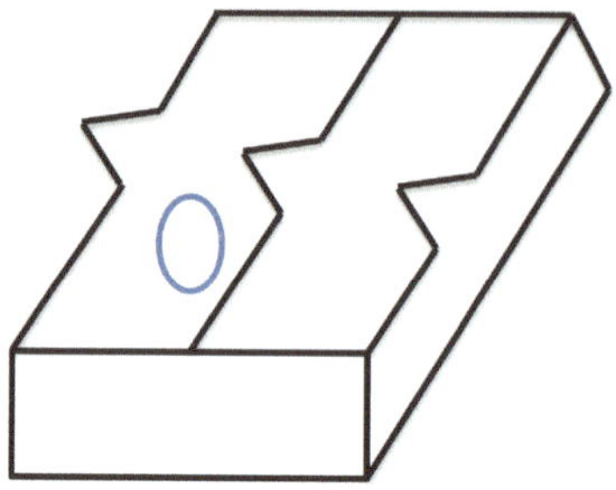

a.  b.  c.  d. 

3. Using Google Earth, search for the following area in Pakistan: **27 50 35.00N 67 10 03.70E**. Zoom out to an eye altitude of approximately 25,000 feet. The inclined layers in these folded rocks can be easily seen. As you view them, in which compass direction are the beds dipping?

a. Northeast (NE) b. Southeast (SE)

c. Northwest (NW) d. Southwest (SW)

12.4 GEOLOGIC STRUCTURES CREATED BY PLASTIC AND BRITTLE DEFORMATION

12.4.1 Folds

Folds are geological formations resulting from the plastic deformation of the Earth's crust. To comprehend the formation of folds, grasp a sheet of paper with one hand at each end. Exert compressive forces by pushing the ends together. You have just formed a fold (bent rock strata). Your paper's movement determined the creation of one of the three primary fold types (Figure 12.11).

Figure 12.11 | The three main fold types, from left to right, are monocline, anticline, and syncline (the anticline and syncline are both displayed in one block).

A **monocline** is a simple fold structure that consists of a bend in otherwise horizontal rock layers. More commonly found are anticlines and synclines. An **anticline** fold is convex up and one in which the layered strata are inclined down and away from the center of the fold (if you drew a line across it, the anticline would resemble a capital letter A). A **syncline** is a concave upward fold in which the layered strata are inclined up (it resembles a smile). Parts of a fold include the axis (the hinge line), the axial plane, and limbs on either side of the axis (Figure 12.12). It is important to note that anticlines do not always represent mountains or high areas and synclines do not always represent basins or low areas. They are simply folded rock layers, and do not necessarily indicate topographic high and low points (see Figure 12.13).

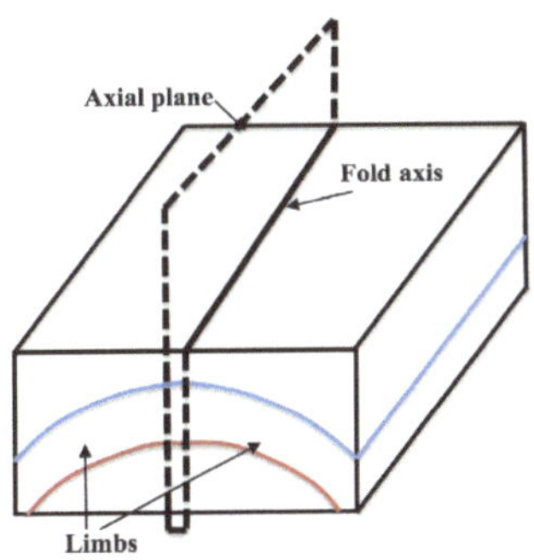

Figure 12.12 | The axial plane and fold axis, along the center of the fold, and corresponding limbs on either side.

Figure 12.13 | This topographic high, along Interstate 68 and US 40 in Maryland, is a syncline.

Folds observed in cross-section look much different from map view. In map view, rather than seeing folds, you will only encounter beds that look like Figure

12.14. To help determine what type of fold you have, it is useful to determine the strike and dip of the beds you encounter. On Figure 12.14, determine the strike and dip for each location marked by an oval. Check yourself against Figure 12.15.

Once rocks are folded and exposed at the Earth's surface, they are subjected to erosion, creating regular patterns. The erosion exposes the interiors of the folds, such that parallel bands of dipping strata can be observed along the fold axis. In an anticline, the oldest rocks are exposed along the fold axis, while it is the youngest rocks exposed at the fold axis in a syncline (Figure 12.16).

Figure 12.14 | A block diagram of an anticline and syncline. Looking from above you see map view, and from the side you observe cross-section view. Determine the strike and dip symbols that should go in the ovals.

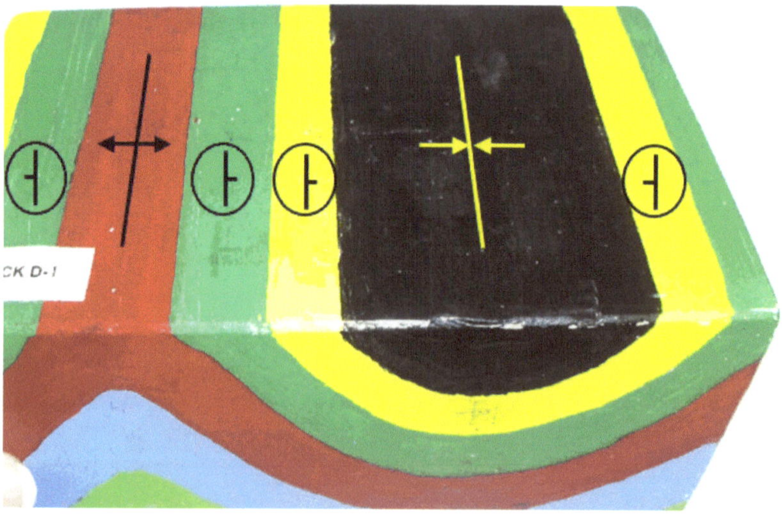

Figure 12.15 | In this block, the strike and dip symbols, along with the symbols for anticline and syncline, have been drawn in for you. Note that on the anticline, the beds dip AWAY from the axis, and the anticline symbol is drawn along the axis, with arrows pointing away from each other. In the syncline, the beds dip TOWARDS each other, with the syncline symbol having arrows that point inwards.

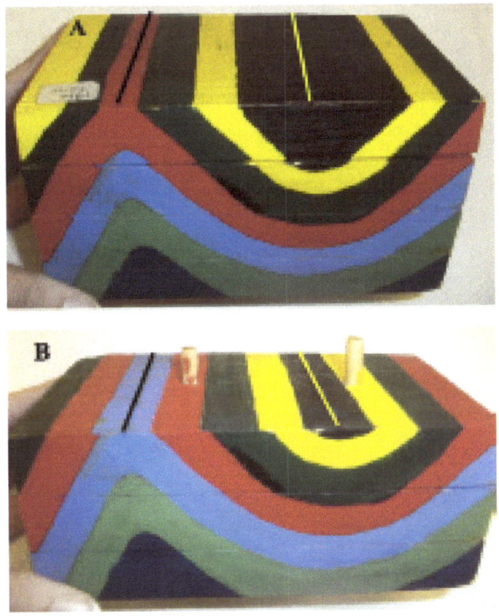

Figure 12.16 | In the top block (A), you view a typical anticline and syncline. Look at the center of the folds. Are the beds that you see in the center of the anticline older or younger than the beds on either side of it? What about for the syncline? In the lower block (B), you see the top portion of the block removed, done so to simulate erosion of rock layers. Note the pale blue bed in the center of the anticline. It is older than the red bed on either side of it (since it is lower in side view—remember older beds are on the bottom). In the syncline, the black bed in the middle is younger than the yellow beds on either side of it.

Table 12.1

Fold Type	Direction of dip of layers	Age of beds at axis
Anticline	Away from axis	Oldest
Syncline	Towards the axis	Youngest

Imagine you take that same sheet of paper that you originally created a fold with. Create a fold again, but rather than holding it horizontally, plunge one end of your paper down into your desk surface. You have now created a plunging fold, which is essentially a tilted fold that creates a V shaped pattern on the surface (Figures 12.17, 12.18, and 12.19). In an anticline, the oldest strata can be found at the center of the V, and the V points in the direction of the plunge of the fold axis. In a syncline, the youngest strata are found at the center of the V, and the V points in the opposite direction of the plunge of the fold axis.

Similar to anticlines and synclines are domes and basins, which are basically the circular (or elliptical) equivalent of those folds. A **dome** is an upfold similar to an anticline, and a **basin** is an area where the rocks are inclined downward towards the center, similar to a syncline (Figure 12.20). The key to identifying these structures is similar to identifying the folds. In a dome, the oldest rocks are exposed at the center, and rocks dip away from this central point. In a basin, the youngest rocks are in the center, and the rocks dip inward towards the center.

Figure 12.17 | Both blocks are plunging folds. Note the V shape created in map view in a plunging fold. The fold on the left block is a plunging anticline (observe the end of the block to determine this) and the fold on the right block is a plunging syncline.

Figure 12.18 | Note this side view of a plunging anticline, to better see how the rock layers plunge into the Earth.

Figure 12.19 | These blocks depict map view for both plunging and non-plunging folds. The top block is a plunging fold, with the characteristic V-shape. The lower block shows map view of non-plunging folds.

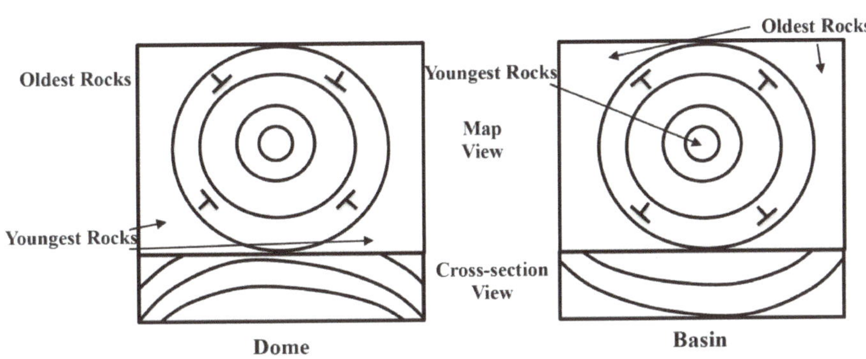

Figure 12.20 | The dome has older rocks in the center, with rocks dipping away from this point, while the basin has rocks dipping inwards and the youngest rocks in the center.

12.4.2 Faults

As rocks undergo brittle deformation, they may produce cracks in the rocks. If no appreciable displacement has occurred along these cracks, they are called joints. If appreciable displacement does occur, they are referred to as faults. We will first examine **dip-slip faults**, in which movement along the fault is either up or down. The two masses of rock that are cut by a fault are termed the fault blocks (Figure

12.21). The type of fault is determined by the direction that the fault blocks have moved.

Fault block movement is described based on the movement of the hanging wall, the fault block located above the fault plane. The other fault block, located beneath the fault plane, is called the foot wall. The term hanging wall comes from the idea that if a miner was climbing along the fault plane, she would hang her lantern above her head, along the hanging wall. Alternately, you can draw a stick figure straight up and down across the fault plane. Its head will be on the hanging wall and its feet will be on the foot wall (Figure 12.22).

Figure 12.21 | Two fault blocks. The fault is the break in the block that separates the two fault blocks.

When extensional forces are applied to the fault blocks, the hanging wall will move down, creating what is called a **normal fault** (an easy way to remember this is the phrase "It's normal to fall down"). As this happens, crust is stretched out and lengthened (Figure 12.23).

When compressional forces are applied to the fault blocks, the hanging wall will move up, creating a **reverse fault**. This causes the crust to shorten in the area (Figure 12.24). A special type of reverse fault is a **thrust fault**. It is a low angle reverse fault (with a dip angle of less than 45o), and has a much thinner hanging wall.

Figure 12.22 | In this image, the head of the stick figure is on the hanging wall (in mauve) and the feet of the stick figure are on the foot wall (in blue).

Figure 12.23 | The hanging wall, on the right, has moved down relative to the footwall, resulting in a normal fault. Notice how close together the cross symbols were in Figure 12.21, compared to this figure, evidence for the lengthening of the block (along red line).

Figure 12.24 | The hanging wall, on the right, has moved up relative to the footwall, resulting in a reverse fault. Note how much closer the cross symbols have become when compared to fi 2 1, evidence for the shortening of the block.

Figure 12.25 | An example of a normal fault near Somerset, UK. Note the fault line in red, and the hanging wall sediments (on the left side) that have moved downward.

Table 12.2

Fault Type	Type of Force	Direction HW moved	Length of Block
Normal	Extensional	Down	Lengthened
Reverse	Compressional	Up	Shortened

Tensional forces acting over a region can produce normal faults that result in landforms known as horst and grabens. In **horst and grabens**, the graben is the crustal block that downdrops, and is surrounded by two horsts, the relatively uplifted crustal blocks (Figure 12.26). This terrain is typical of the Basin and Range of the western United States. The previous faults we observed were dip-slip faults, where movement occurred parallel with the fault's dip. In a **strike-slip fault**, horizontal motion occurs (in the direction of strike, hence the name), with blocks on opposite sides of a fault sliding past each other due to shear forces.

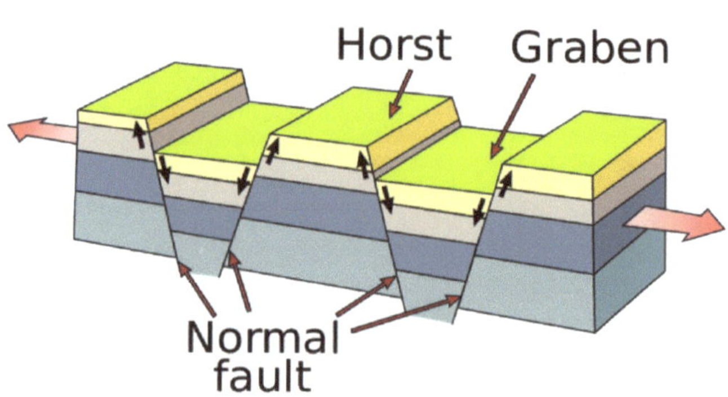

Figure 12.26 | This figure depicts an area that has been stretched by tensional forces, resulting in numerous normal faults and horst and graben landforms.

The classic example of a strike-slip fault is the San Andreas Fault in California (Figure 12.27). These faults can be furthered classified as right-lateral or left-lateral. To determine this, an observer would stand along one side of the fault, looking across at the opposite fault block. If that fault block appears to have moved right, it is right-lateral; if it has moved left, it is left-lateral. Figure 12.28 provides all three fault types for your review.

Figure 12.27 | A block diagram of the San Andreas Fault, a right-lateral strike-slip fault.

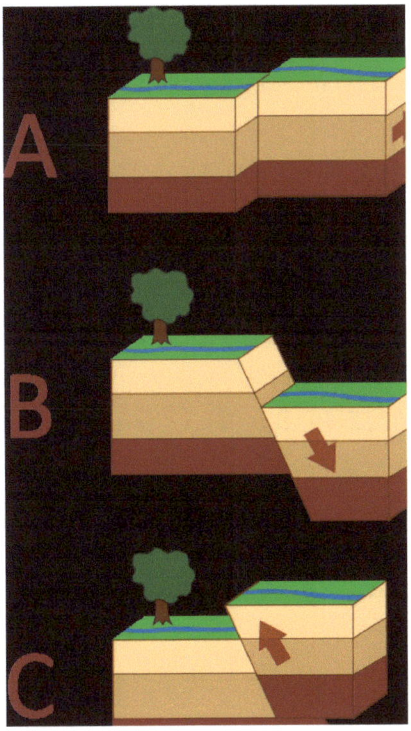

Figure 12.28 | The three types of faults discussed in this lab. A is a strike-slip fault, B is a normal fault, and C is a reverse fault.

12.5 GEOLOGIC MAPS AND CROSS-SECTIONS

A **geologic map** uses lines, symbols, and colors, to include information about the nature and distribution of rock units within an area. It includes a base map, over which information about geologic contacts and strikes and dips are included. Geologists make these maps by careful field observations at numerous outcrops (exposed rocks at the Earth's surface) throughout the mapping area. At each outcrop, geologists record information such as rock type, strike and dip of the rock layers, and relative age data. Geologic maps take practice to understand, as three-dimensional features (such as folds) are displayed on a two-dimensional surface. Remember that a geologic map will be seen in map view, as we learned about earlier with the block diagrams. Geologists use information about rocks that are exposed to visualize the unseen rocks beneath the surface, enabling them to complete the cross-sectional views we observed in the blocks. You will examine a geologic map for an area in Georgia and construct a **geologic cross-section.** A geologic cross-section shows geologic features from the side view. They are similar to the topographic profiles that you have already created, but include additional information about the rocks present. In order to construct a geologic cross-section, obey the following steps:

1. Observe the geologic map given. Pay close attention to any strike and dip symbols, geologic contacts, and ages of the rock types.
2. Take a scratch sheet of paper. Line it up along the line provided across the cross-section.
3. At each geologic contact, make a mark on the scratch paper. Position the marks in the direction you believe the rocks are dipping (To determine this, use strike and dip symbols. If they are not provided, use the Rule of V's or the ages given to help determine the geologic structure).
4. Transfer the marks from your paper to a provided diagram.
5. Sketch in the structure, paying careful attention to dip angles (if provided). Structures may be drawn in with a dotted line above the Earth's surface to indicate rocks that were formerly present but that have since been eroded.

There are some helpful hints to remember when constructing a cross-section:

1. Anticlines – these folds have the oldest beds in the middle, with beds dipping away from the axis. Plunging anticlines plunge towards the closed end of the V.
2. Synclines – these folds have the youngest beds in the middle, with beds dipping towards the axis. Plunging synclines plunge towards the open end of the V.

3. As streams intersect dipping beds of rock, they will cut V shapes in the direction of dip.

Figure 12.29 provides an example of a simple geologic cross-section, based off ages of the rock units.

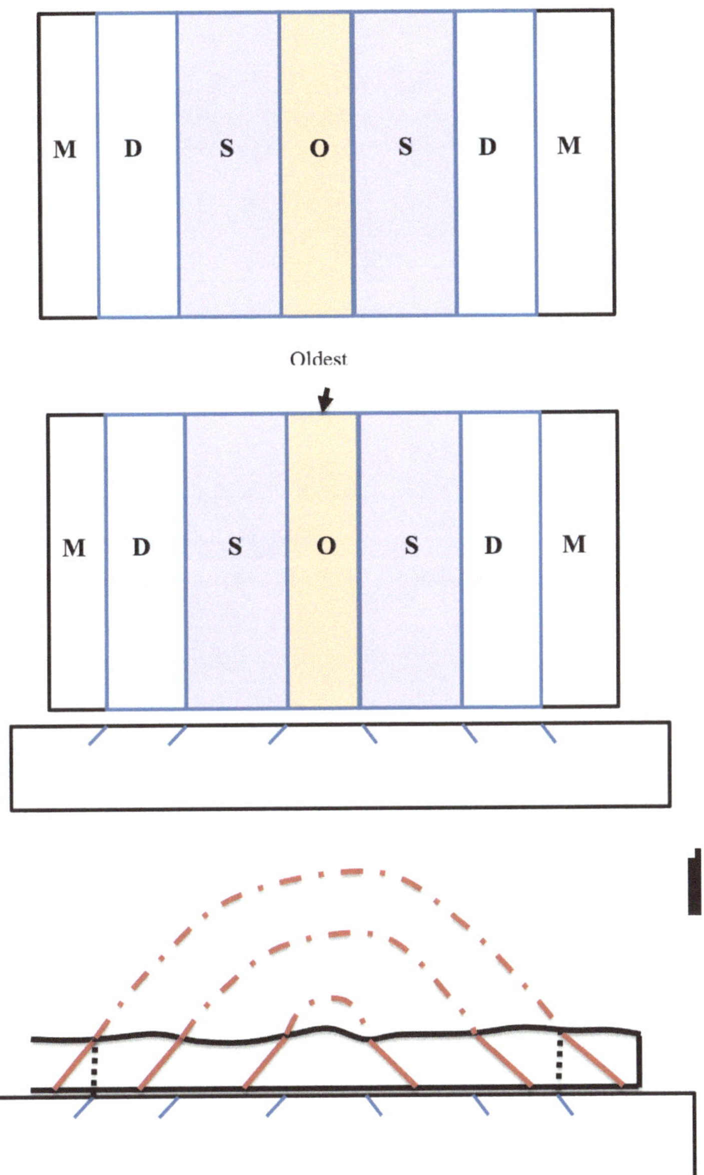

Figure 12.29 | The top figure displays the geologic map, with different rock units of different colors, and age information given (see Figure 12.30).

In the second figure, the scratch sheet of paper is stretched along the bottom of the cross-section, and a mark is made at each geologic contact. Since the oldest bed is in the middle, this indicates that this structure is an anticline, so beds have been drawn so that they dip away from each other. In the bottom image, the scratch paper has been superimposed and contacts have been drawn in (as indicated by the dashed lines on each end mark). Beneath the surface, contacts have been drawn in using a solid line. The beds above the surface that have since eroded have been drawn in with a dashed line.

Age of Rocks	Geologic Age Symbol	Geologic Time Period
Youngest	Q	Quaternary
	T	Tertiary
	K	Cretaceous
	J	Jurassic
	Tr	Triassic
	P	Permian
	lP	Pennsylvanian
	M	Mississippian
	D	Devonian
	S	Silurian
	O	Ordovician
	-C	Cambrian
Oldest	p-C	PreCambrian

Figure 12.30 | Guide to rock ages.

Map Symbol	Explanation
	Strike & Dip
	Vertical strata
	Horizontal strata
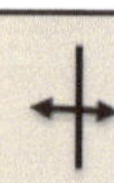	Anticline axis
	Syncline axis
	Plunging anticline axis

	Plunging syncline axis
	Strike-slip fault

Figure 12.31 | Guide to common map symbols.

12.6 PRACTICE WITH BLOCK DIAGRAMS

For practice, complete the questions below about block diagrams. A key is provided after this section to check your work.

For each diagram, draw in the geological contacts on each side of the block. Add strike and dip symbols, and other symbols to document geologic features (like direction of movement on faults). Also state the name of the geology feature in the diagram.

1.
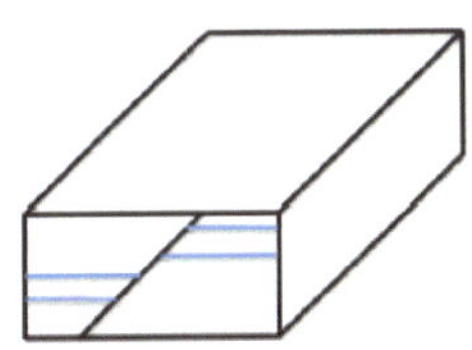

2.
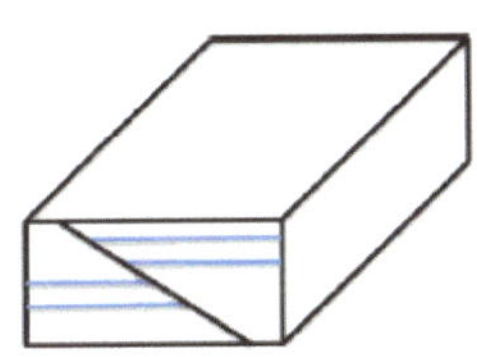

3.
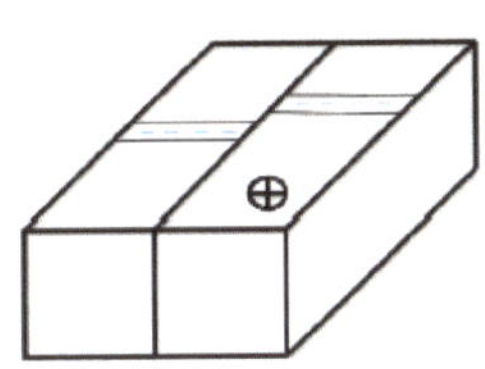

4.
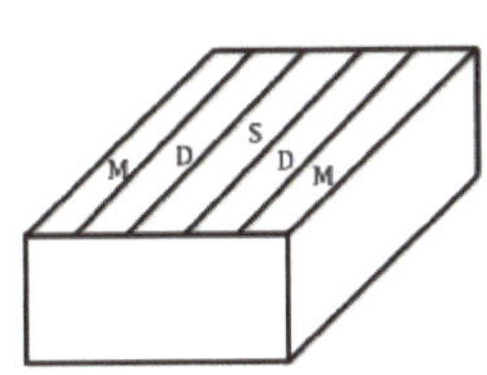

5.
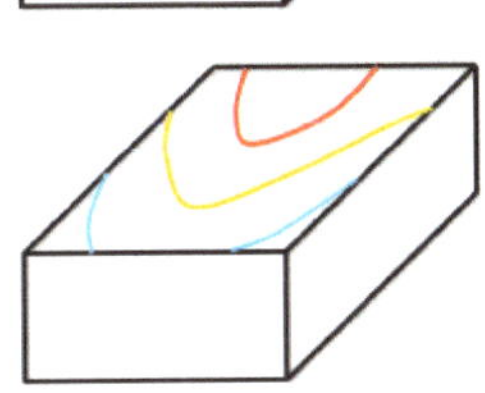

12.7 ANSWERS TO PRACTICE WITH BLOCK DIAGRAMS

1. NORMAL FAULT

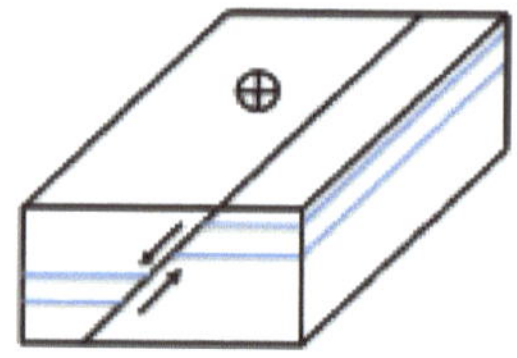

2. REVERSE FAULT

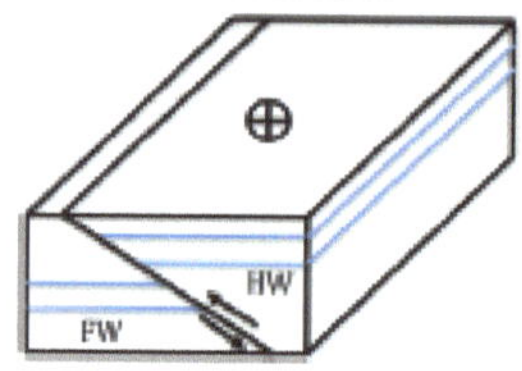

3. STRIKE-SLIP FAULT

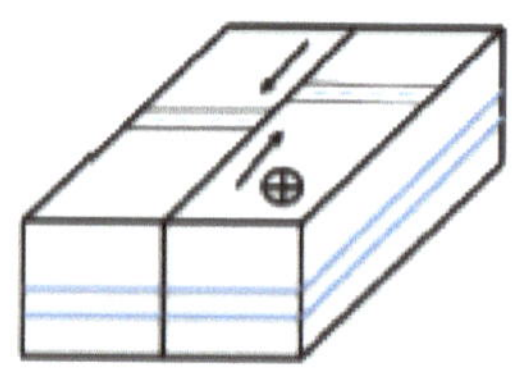

4. ANTICLINE

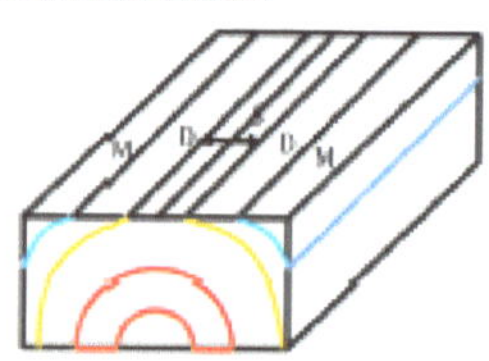

5. PLUNGING SYNCLINE

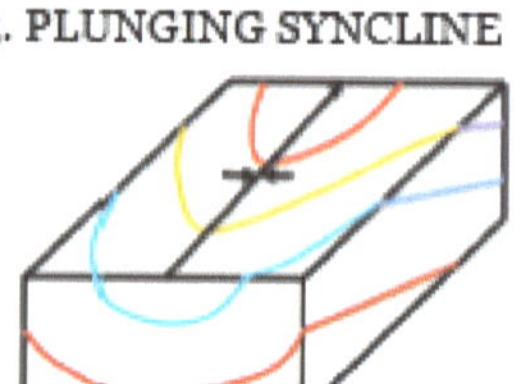

12.7.1 LAB EXERCISE

This Lab Assignment must be mailed to your Instructor. There is no online assessment for the Crustal Deformation Lab.

Complete the entire assignment and mail to your instructor postmarked by the assessment deadline. You should make an extra copy to practice on and mail in a clean and neat version for grading. Make sure to include your name and staple all of the pages together. It is a good idea to make a copy of what you mail in, just in case it gets lost in the mail. **For several parts of this lab there are multiple interpretations! Ask if you have questions.**

Part A (6 pts) Circle the correct answer to the following questions.

1. For the following diagram, determine the correct map symbol that would go in the oval box.

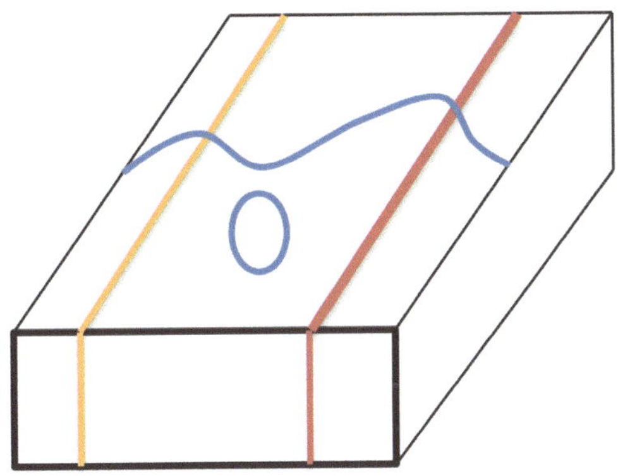

a. ⊢ b. ⊣ c. + d. ⊕

2. For the following diagram, determine the correct map symbol that would go in the oval box.

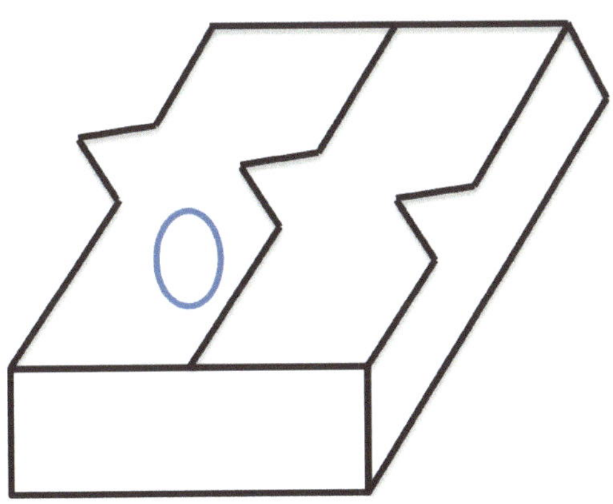

a. ⊢ b. ⊣ c. + d. ⊕

3. Using Google Earth, search for the following area in Pakistan: **27 50 35.00N 67 10 03.70E**. Zoom out to an eye altitude of approximately 25,000 feet. The inclined layers in these folded rocks can be easily seen. As you view them, in which compass direction are the beds dipping?

   a. Northeast (NE) b. Southeast (SE) c. Northwest (NW) d. Southwest (SW)

Part B (24 pts)

For each of the following block diagrams complete the following: 1Complete the diagram drawing in geological contacts on each side of the block; 2Add symbols indicating the strike and dip of each geological layer as well as symbols documenting any other geological features (include the direction of movement for any faults); 3In the space provided under the block diagram write the specific name of the geology feature in the block diagram. Please note that in several of the blocks the ages of the layers are provided.

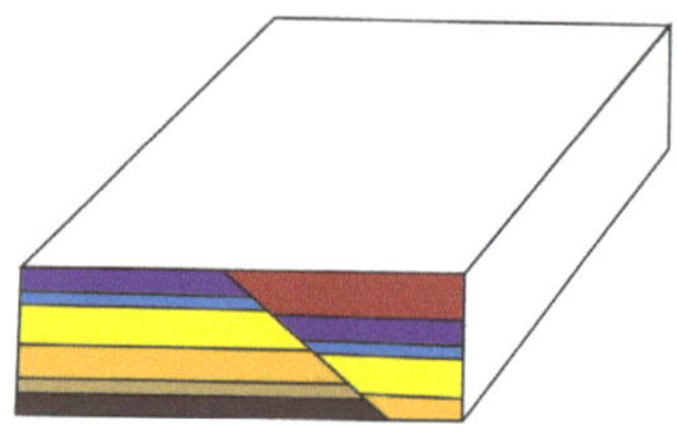

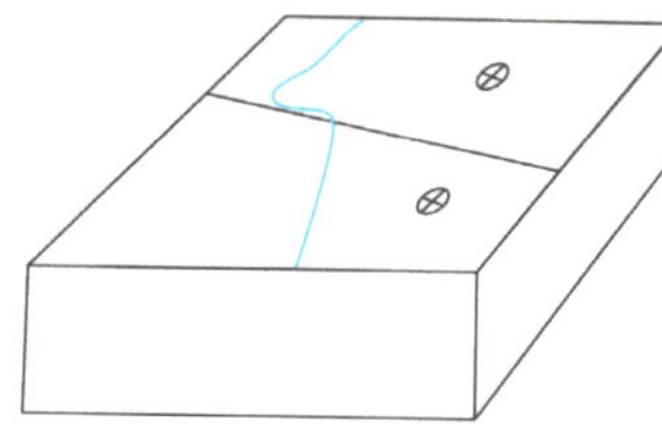

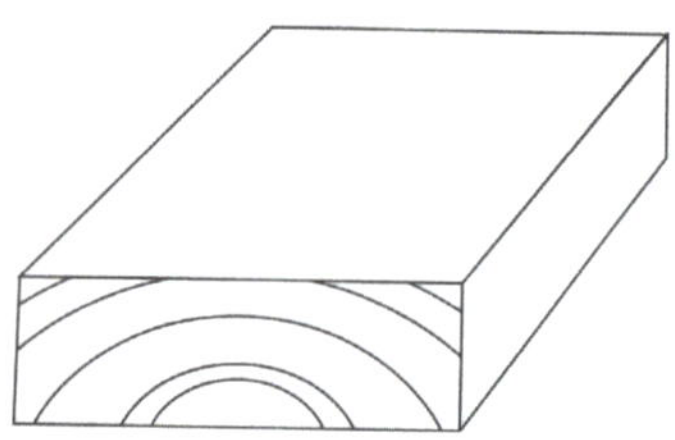

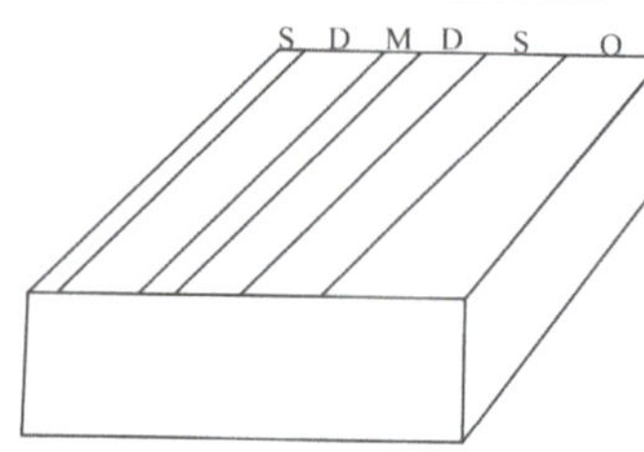

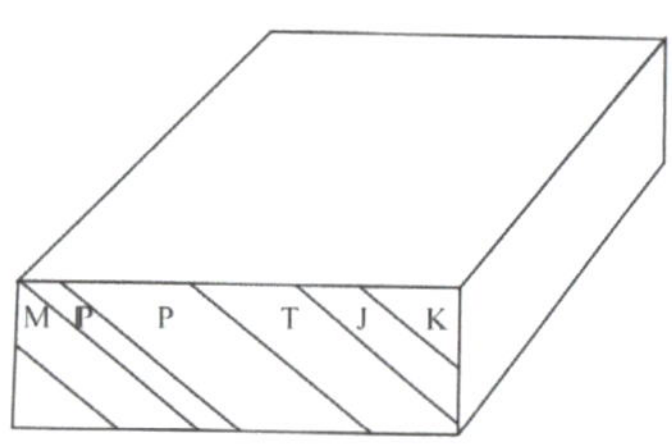

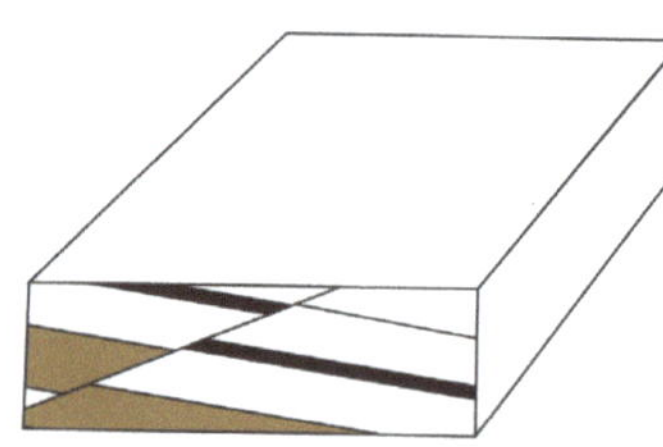

Part C (20 pts)

At the end of the lab there are two full-page block diagrams. Cut along the dashed lines and fold along the solid lines to examine the block in three dimensions. For each of the block diagrams complete the following: 1Complete the diagram drawing in geological contacts on each side of the block; 2Add symbols indicating the strike and dip of each geological layer as well as symbols documenting any other geological features; 3Identify the geologic structure presented on the block and write the name of the structure on the top surface of the block. Please unfold the blocks flat to mail in to your instructor.

Part D (24 pts)

The geological map on the following page is from the Paleozoic rocks in the Northwest Georgia Mountains. Please refer to the key to rocks and their ages present on the map in order to answer the questions. Note that the rock key is in chronological order with the oldest rocks on the bottom (Conasauga shale) and the youngest rocks on top (Pennsylvanian Fm).

- Cpv Pennsylvanian Fm
- Cbp Mississippian Fm
- Dc Chattanooga shale
- Srm Red Mountain Fm
- Om Upper Chickamauga Ls
- Osr Lower Chickamauga (Stones River)
- On Newala limestone (Upper Knox)
- OЄk Knox dolomite
- Єc Conasauga shale
- Єr Rome Fm
- Єs Shady dolomite

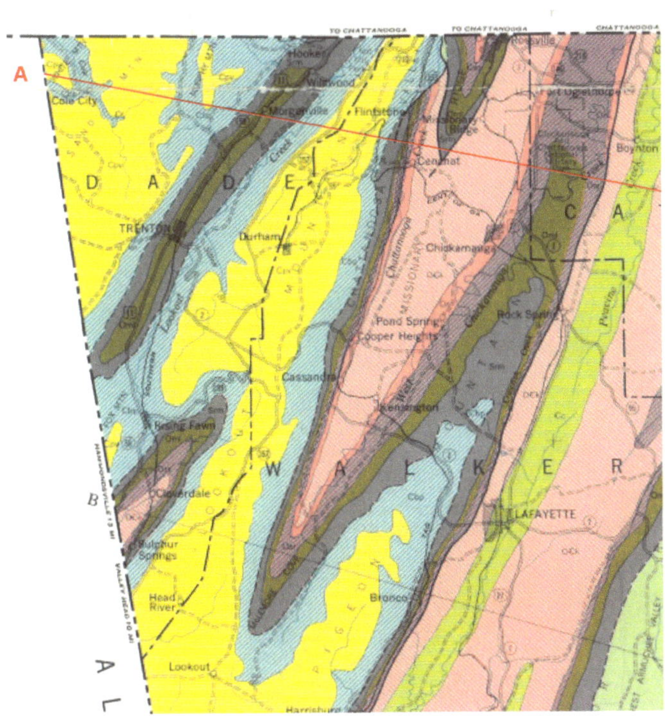

4. (6 pts) Examine the map and the ages of the rock layers. Based on your interpretation, label the structures and add appropriate symbols to the map to show the axis of the folds. What kind of folds are present in this area? Explain how you came to this conclusion.

5. (6 pts) Along the red line starting at point 'A' there are four faults indicated by bold black lines (they are also marked on the cross section in Part C). Examine the age of the rocks and the faults themselves. What type of faults are these? Explain how you came to this conclusion.

6. (12 pts) Complete the following geological cross section based on your understanding of the geological structures present in the area.

A

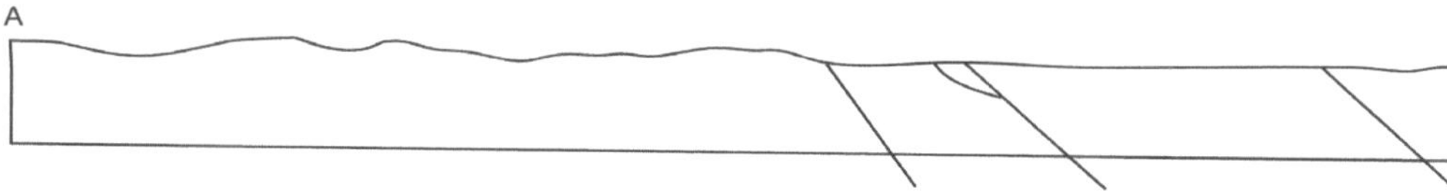

Part E Google Earth. (26 pts)

7. Search for **38 04 36.00N 109 55 26.61 W** and zoom to an eye altitude of 20,000 ft, make sure to note your geographic location. (hint: re-read the section in the course content unit on Faults and in this lab manual)

    a. You are looking at the surface expression of many fault lines (fault scarp) with the hanging wall occupying the valleys and the foot wall representing the plateaus, what kind of faults are they?

    b. What would we call these **paired** features?

    c. What type of stress created these features?

    d. The amount of displacement (extension or shortening) in the area can be calculated by the following equation based on the angle of the fault surface:

Amount of horizontal displacement = Vertical displacement along the fault surface (ft)/5.67

Measure the change in elevation of one of these features to determine the amount of extension in the area (Show your work, but feel free to use a calculator).

    e. How many of these faults would you need to make this area (Utah) a mile longer? Remember there are 5,280 feet in a mile.

8. Search for **22 48 45.63S 117 20 12.85E** and zoom to an eye altitude of 25 miles. This feature is caused by the folding of rocks.

    a. Imagine you are a geologist trying to determine what is going on in the area. How could you tell if this large bullseye structure is a dome or a basin (State whether this is a dome or basin, and give two different ways to tell them apart)?

    b. Examine the feature ~10 miles southwest of the previous feature. Is this fold horizontal or plunging, how can you tell?

{Part C-1)

c.

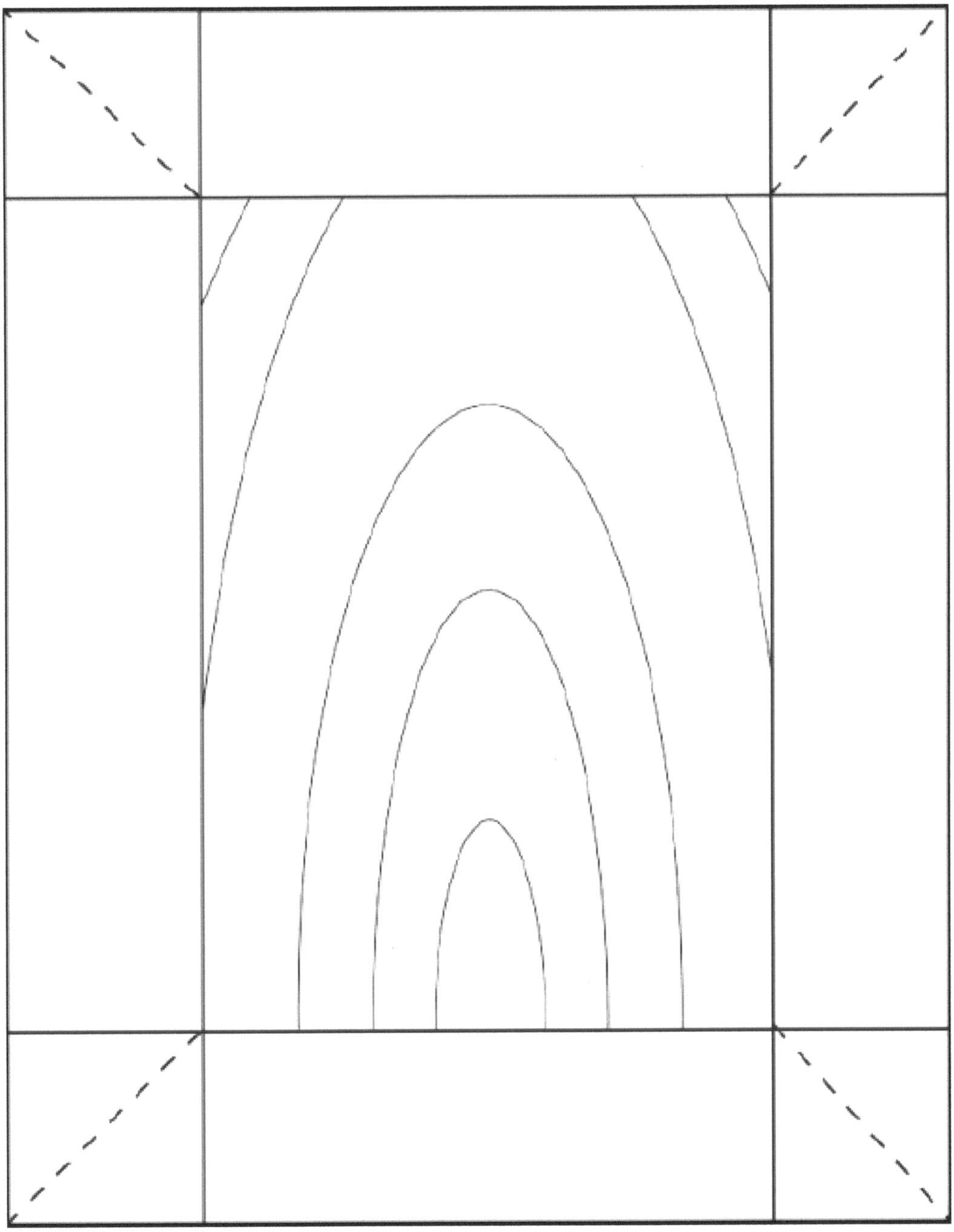

(Part C-2)

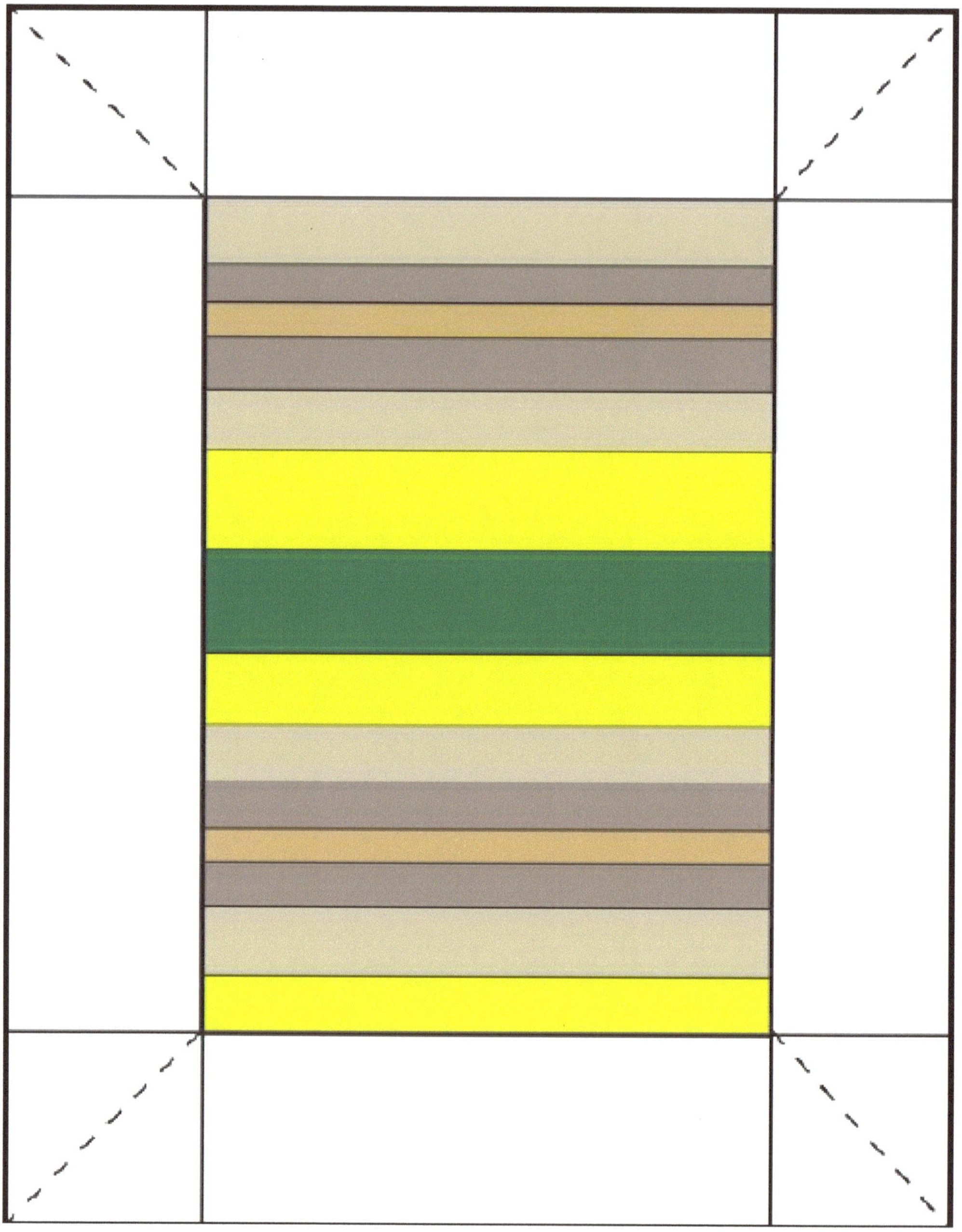

13 Chapter | EARTHQUAKES

13.1 INTRODUCTION: AN OVERVIEW

It marked the most lethal day in the annals of Mt. Everest. On April 25, 2015, a 7.8 magnitude earthquake struck Nepal. This precipitated an avalanche that resulted in the fatalities of 19 climbers on Mt. Everest. In Nepal, at than 8,800 individuals perished, with numerous others injured and rendered homeless. Numerous aftershocks have transpired since the first earthquake (Figure 13.1).

This region has a history of earthquakes. A comparable death toll was recorded after a 1934 earthquake, alongside other lesser seismic events throughout history. A 1833 earthquake of comparable size caused less than 500 fatalities, perhaps attributable to two significant foreshocks that prompted most occupants to evacuate outdoors in fright, so enhancing their safety. Globally, this century has witnessed far more lethal and powerful earthquakes, such as the Haiti earthquake in 2010, which resulted in 316,000 fatalities, and the Sumatra earthquake in 2004, which caused 227,000 deaths, both attributed to ground shaking and subsequent dangers. Earthquakes provide geologists with critical insights into the Earth's interior, as discussed in the Earth's Interior chapter, as well as surface conditions, since most earthquakes transpire at plate borders, as outlined in the Plate Tectonics chapter, Figure 4.8.

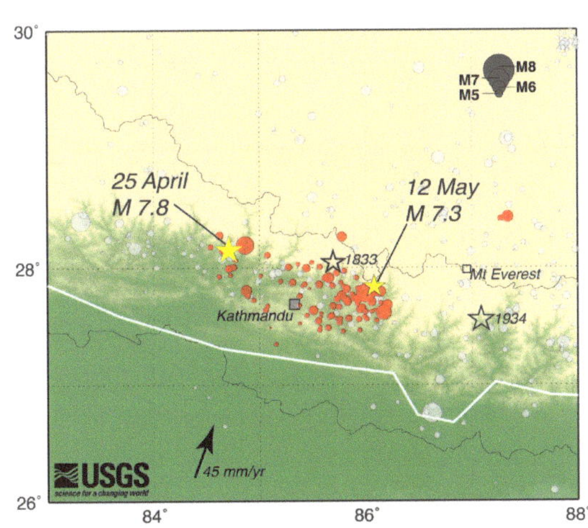

Figure 13.1 | A map of the main earthquake to hit Nepal on April 25, 2015, along with a major aftershock on May 12, and numerous (>100) other aftershocks (in red – note the magnitude scale in the upper right).

13.2 THE EPICENTER, FOCUS, AND WAVES

An earthquake is a like a telegram from the Earth. It sends a message about the conditions beneath the Earth's surface. The shaking or trembling experienced during an earthquake is the result of a rapid release of energy within the Earth, usually as a result of movement along geologic faults. Think back to the strike-slip fault from the Crustal Deformation chapter. Rocks on either side of the fault are sliding past each other. As they move in opposite directions, the rocks become deformed, as they will bend slightly and build up pressure. Eventually they will reach a breaking point. Once the strength of the rock has been exceeded, the rocks will snap back to their normal shape, releasing all that stored energy as an earthquake. The more energy that has been stored, the larger the earthquake is. Remember the stress-strain diagram from Crustal Deformation.

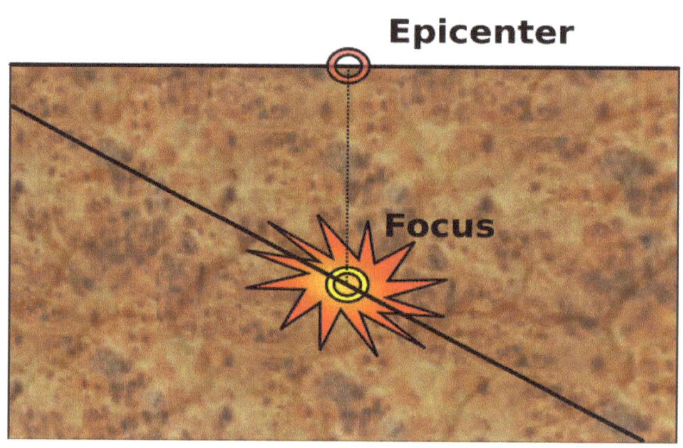

Figure 13.2 | An illustration depicting the focus, where the earthquake originates, and the epicenter, the point on the ground's surface directly above the focus.

When rocks are under too great of a stress, they undergo brittle failure (the earthquake). The strength of the rock has been exceeded at this point.

Earthquakes originate at a point called the **focus** (plural foci). From this point, energy travels outward in different types of waves. The place on the Earth's surface directly above the focus is called the **epicenter** (Figure 13.2). Earthquake foci may be shallow (less than 70 km from Earth's surface) to deep (greater than 300 km deep), though shallow to intermediate depths are much more common. Earthquake frequency and depth are related to plate boundaries. The vast majority (95%) of earthquakes occur along a plate boundary, with shallow focus earthquakes tending to occur at divergent and transform plate boundaries, and shallow to intermediate to deep focus earthquakes occurring at convergent boundaries (along the subducting plate). The earthquakes associated with convergent boundaries occur along Wadati-Benioff zones, or simply **Benioff zones**, areas of dipping seismicity along the subducting plate (Figure 13.3).

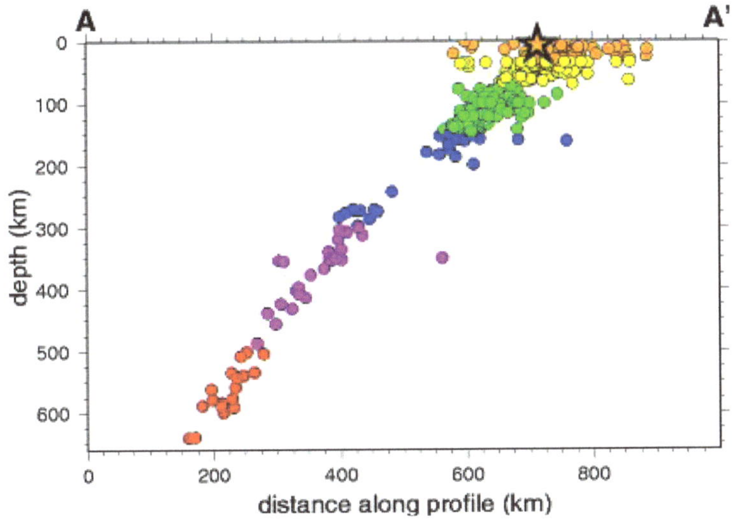

Figure 13.3 | This is a seismicity cross-section, taken along the subducting plate in an ocean-ocean convergent boundary at the Kuril Islands, located just northeast of Japan. Foci are located in the descending slab. Only brittle substances (like the lithosphere) can generate earthquakes, so this must be the subducting slab. The star represents the location of an 8.3 magnitude earthquake that occurred on 11/15/06.

As an earthquake occurs, two different types of waves are produced; **body waves**, so termed because they travel through the body of the Earth, and **surface waves** that travel along the Earth's surface (Figure 13.4). There are two types of body waves. **P-waves**, or primary waves, are compressional waves that move back and forth, similar to the action of an accordion. As the wave passes, the atoms in the material it is travelling through are being compressed and stretched. Movement is compressional parallel to the direction of wave propagation, which makes P-waves the fastest of the seismic waves. These waves can travel through solids, liquids, and gases, because all materials can be compressed to some degree. **S-waves**, or secondary waves, are shear waves that move material in a direction perpendicular to the direction of travel. S-waves can only travel through solids, and are slower than P-waves. A similar motion to S-wave motion can be created by two people holding a rope, with one snapping the rope quickly. Alternately, you can also think of this wave movement similar to the wave created by fans in a stadium that stand up and sit down. Body waves are responsible for the jerking and shaking motions felt during an earthquake.

Surface waves are slower than body waves, and tend to produce more rolling sensations to those experiencing an earthquake, similar to being in a boat on the sea. Because surface waves are located at the ground's surface where humans (and their structures) are located, and because they move so slowly, which bunches them up and increases their amplitude, they are the most damaging of seismic waves. **Love waves** are the faster surface waves, and they move material back and forth in a horizontal plane that is perpendicular to the direction of wave travel (see Figure 13.4). Buildings do not handle this type of movement well, and Love waves may be responsible for considerable damage to structures. **Rayleigh waves** make the Earth's surface move in an elliptical motion, similar to the movement in a sea wave. This results in ground movement that is up and down and side-to-side.

13.3 SEISMOLOGY

Humans have experienced earthquakes throughout their existence on Earth, although most ancient cultures created mythology to elucidate them, often imagining colossal animals within the Earth whose movements caused the tremors. Seismology, the study of earthquakes, advanced significantly with the invention of instruments capable of detecting seismic activity; the seismograph, in particular, can record even the most minute vibrations of the Earth (Figure 13.5). A conventional seismograph comprises a mass suspended by a string from a frame that oscillates in response to the Earth's surface movements. A revolving drum is affixed to the frame, with a pen connected to the mass, enabling the recording of relative motion on a seismogram. During an earthquake, it is the frame anchored to the ground that shifts, while the hanging mass typically remains stationary due to inertia, which is the propensity of an object to remain at rest and oppose motion.

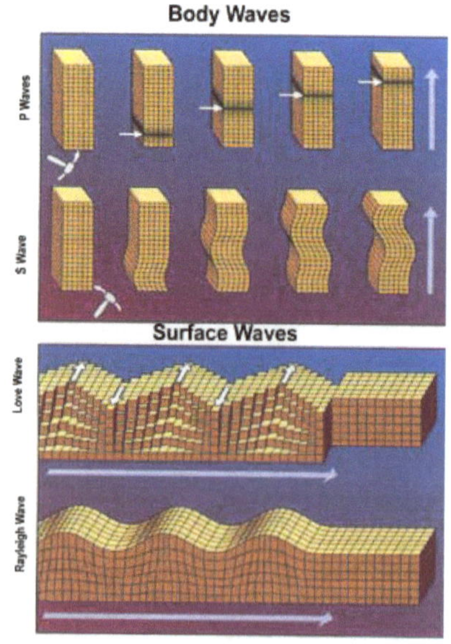

Figure 13.4 | The different types of seismic waves. Body waves, in the upper portion of the figure, consist of P-waves and S-waves. The P-wave motion is compressional. The hammer on the left starts the wave moving. The arrow on the right shows the general direction of the wave. In the S-wave, the motion is undulating. Surface waves are depicted in the lower portion of the chart. Love

waves move similarly to S-waves, resulting in horizontal shifting of the Earth's surface. Rayleigh waves are surface waves that travel much like a wave along the water's surface.

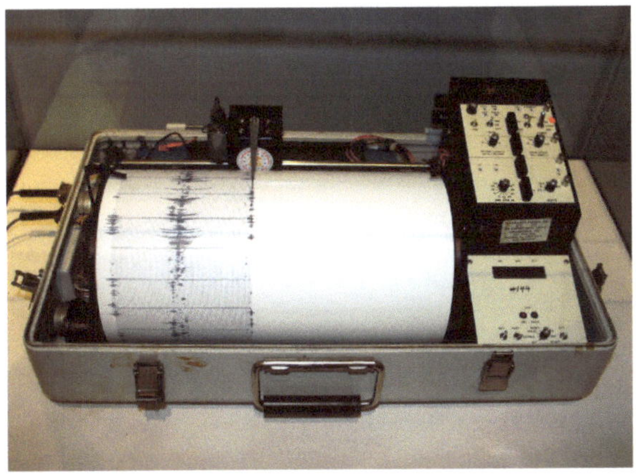

Figure 13.5 | A seismograph and the seismogram it produces.

13.3.1 How Are Earthquakes Measured?

The tragic consequences of earthquakes can be measured in many ways, like death tolls or force of ground shaking. Two measures in particular are commonly used. One is a qualitative measure of the damage inflicted by the earthquake, and it is referred to as **intensity**. The second is a quantitative measure of the energy released by the earthquake, termed **magnitude**. Both measures provide meaningful data.

13.3.2 Earthquake Intensity

Intensity measurements take into account both the damage incurred due to the quake and the way that people respond to it. The Modified Mercalli Intensity Scale (Figure 13.6) is the most widely used scale to measure earthquake intensities. This scale has values that range from Roman numerals I to XII which characterize the damage observed and people's reactions to it. Data for this scale is often collected right after an earthquake by having the local population answer questions about the damage they see and what happened during the quake. This information can then be pooled to create an intensity map, which creates colored zones based on the information collected (Figure 13.7). These maps are frequently used by the insurance industry.

Intensity	Characteristics
I	Shaking not felt under normal circumstances.
II	Shaking felt only by those at rest, mostly along upper floors in buildings.
III	Weak shaking felt noticeably by people indoors. Many do not recognize this as an earthquake. Vibrations similar to a large vehicle passing by.

IV	Light shaking felt indoors by many, outside by few. At night, some were awakened. Dishes, doors, and windows disturbed; walls cracked. Sensation like heavy truck hitting a building. Cars rock noticeably.
V	Moderate shaking felt by most; many awakened. Some dishes and windows broken. Unstable objects overturned.
VI	Strong shaking felt by all, with many frightened. Heavy furniture may move, and plaster breaks. Damage is slight.
VII	Very strong shaking sends all outdoors. Well-designed buildings sustain minimal damage; slight-moderate damage in ordinary buildings; considerable damage in poorly built structures.
VIII	Severe shaking. Well-designed buildings sustain slight damage; considerable damage in ordinary buildings; great damage in poorly built structures.
IX	Violent shaking. Well-designed buildings sustain considerable damage; buildings are shifted off foundations, with some partial collapse. Underground pipes are broken.
X	Extreme shaking. Some well-built wooden structures are destroyed; most masonry and frame structures are destroyed. Landslides considerable.
XI	Few structures are left standing. Bridges are destroyed, and large cracks open in the ground.
XII	Total damage. Objects thrown upward in the air.

Figure 13.6 | (Above) An abbreviated table of the Modified Mercalli Intensity Scale. Intensity for a particular earthquake is determined by the maximum damage incurred.

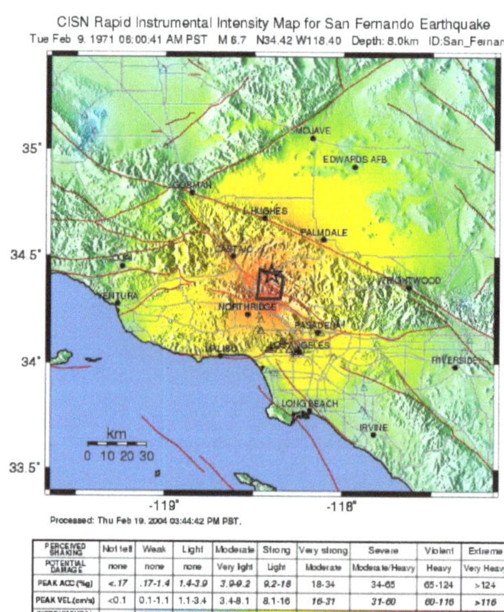

Figure 13.7 | (Right) An intensity map for the San Fernando earthquake in southern California on 2/9/76. Notice that near the epicenter (marked by a star), the intensity was extreme.

13.3.3 Earthquake Magnitude

Another way to classify an earthquake is by the energy released during the event; this is referred to as the magnitude of the earthquake. While magnitude has been measured using the Richter scale, as the frequency of earthquake measurements around the world increased, it was realized that the Richter magnitude scale was not valid for all earthquakes (it is not accurate for large magnitude earthquakes). A new scale called the Moment Magnitude Intensity Scale was developed, which maintains a similar scale to the Richter scale. This scale estimates the total energy released by an earthquake and can be used to characterize earthquakes of all sizes throughout the world. The magnitude is based on the seismic moment (estimated based on ground motions recorded on a seismogram), which is a product of the distance a fault moved and the force required to move it. This scale works particularly well with larger earthquakes and has been adopted by the United States Geological Survey. Magnitude is based on a logarithmic scale, which means for each whole number that you increase, the amplitude of the ground motion recorded by a seismograph increases by 10 and the energy released increases by 101.5, rather than one (so that a 3 magnitude quake results in ten times the ground shaking as a 2 magnitude quake; a magnitude 4 quake has 102 or 100 times the level of ground shaking as a 2 magnitude quake (releasing 103 or 1000 times as much energy). For a rough comparison of magnitude scale to intensity, see Figure 13.8. Why is it necessary to have more than one type of scale? The magnitude scale allows for worldwide characterization of any earthquake event, while the intensity scale does not. With an intensity scale, a IV in one location could be ranked a II or III in another location, based off of building construction (ex. poorly constructed buildings will suffer more damage in the same magnitude earthquake as those built with stronger construction).

Magnitude	Typical Maximum Modified Mercalli Intensity
1.0 – 2.9	I
3.0 – 3.9	II – III
4.0 – 4.9	IV – V
5.0 – 5.9	VI – VII
6.0 – 6.9	VII – IX
7.0 and above	VIII or above

Figure 13.8 | A comparison of magnitude versus intensity scales for earthquakes.

13.4 LOCATING AN EARTHQUAKE EPICENTER

During an earthquake, seismic waves are sent all over the globe. Though they may weaken with distance, seismographs are sensitive enough to still detect these waves. In order to determine the location of an earthquake epicenter, seismographs from at least three different places are needed for a particular event. In Figure 13.9, there is an example seismogram from a station that includes a minor earthquake.

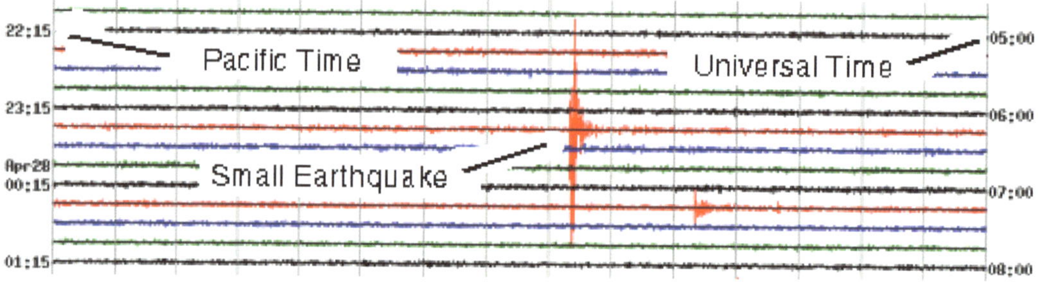

Figure 13.9 | This seismogram is read from left to right and top to bottom. Note the small earthquake that is marked, and the resulting change in wave amplitude at that point.

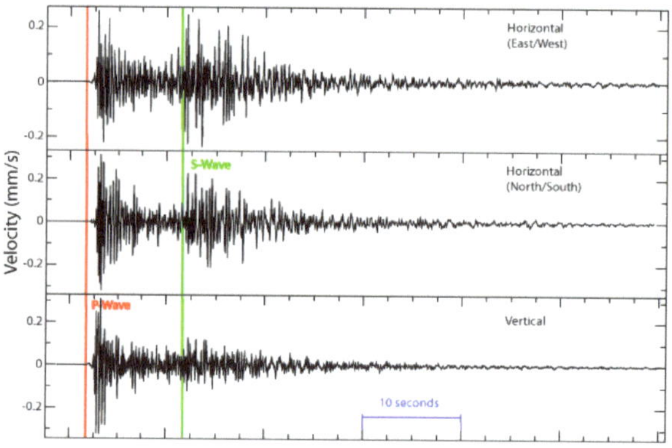

Figure 13.10 | An example seismogram with the arrival of P and S waves included. Note how the arrival of waves is marked by an increase in the wave height (known as amplitude) and by more tightly packed waves. This example does not include time along the bottom, but those in the lab exercise will.

Once three seismographs have been located, find the time interval between the arrival of the P-wave and the arrival of the S-wave. First determine the P-wave arrival, and read down to the bottom of the seismogram to note at what time (usually marked in seconds) that the P-wave arrived. Then do the same for the S-wave. The arrival of seismic waves will be recognized by an increase in amplitude – look for a pattern change as lines get taller and more closely spaced (ex. Figure 13.10).

By looking at the time between the arrivals of the Pand S-waves, one can determine the distance to the earthquake from that station, with longer time intervals indicating longer distance. These distances are determined using a travel-time curve, which is a graph of Pand S-wave arrival times (see Figure 13.11).

Though distance to the epicenter can be determined using a travel-time graph, direction cannot be told. A circle with a radius of the distance to the quake can be drawn. The earthquake occurred somewhere along that circle. Triangulation is required to determine exactly where it happened. Three seismographs are needed. A circle is drawn from each of the three different seismograph locations, where the radius of each circle is equal to the distance from that station to the epicenter. The spot where those three circles intersect is the epicenter (Figure 13.12).

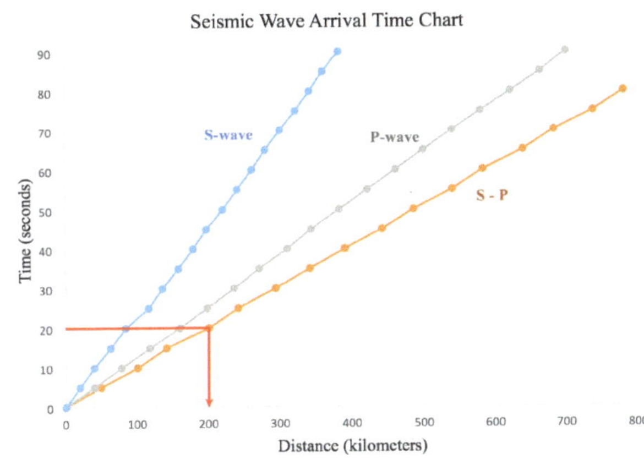

Figure 13.11 | A travel-time graph that includes the arrival of P-waves and S-waves. Note that these curves plot distance versus time, and are calculated based on the fact that the Earth is a sphere. Curves vary with the depth of earthquake because waves behave differently (i.e. their velocities change) with depth and change in material. This particular curve is used for shallow earthquakes (<20 km deep) with stations within 800 km. The S-P curve refers to the difference in time between the arrival of the P-wave and S-wave. If you noted on your seismogram that the P-wave arrived at 10 seconds, and the S-wave arrived at 30 seconds, the difference between arrival times would be 20 seconds. You would read the 20 seconds off the y-axis above to the S-P line, then drop down to determine the distance to the epicenter. In this case, it would be approximately 200 kilometers.

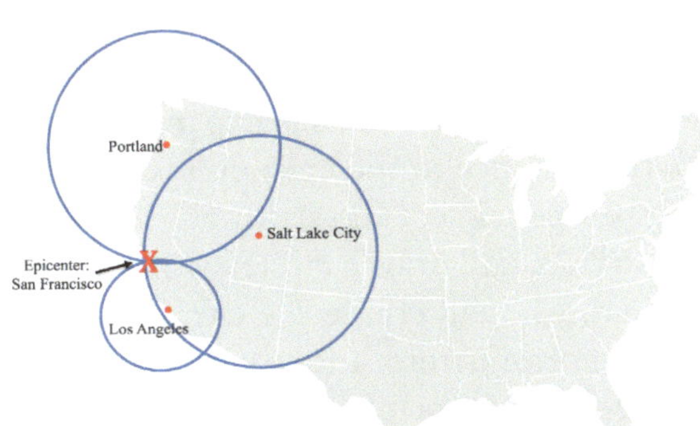

Figure 13.12 | In order to locate this earthquake epicenter, seismograms from Portland, Salt Lake City, and Los Angeles were used. The time between P and

S wave arrivals was calculated, and travel time tables gave a distance. Circles with each distance for its radii were drawn from each station. The one resulting overlap, at San Francisco, was the earthquake epicenter.

13.4.1 LAB EXERCISE

Part A – Locating an Epicenter

You will determine the location of an earthquake epicenter using seismograms from Carrier, Oklahoma, Smith Ranch in Marlow, Oklahoma, and Bolivar Missouri available at the end of this chapter. These are actual seismograms that you will be reading, from an actual event. For each, three different readouts are given, as the seismograph measured in three different axes. You may focus on any of the three readouts for each station, as all will have the same arrival times for each wave. First, determine when the P and S waves arrived, and note these times (remember to look for a pattern change as lines get taller and more closely spaced). Mark both the arrival of the P-wave and S-wave, then using the time scale in seconds, note the time difference between the P and S wave arrivals. Add this to the table below for each of the three seismograms.

Table 13.1

Station	P-wave arrival time (sec)	S-wave arrival time (sec)	Difference between P and S travel times (sec)	Distance to Epicenter from Station (km)
Carrier, OK				
Marlow, OK				
Bolivar, MO				

The different between the P and S wave arrivals will be used to determine the distances to the epicenter from each station using Figure 13.11. Make sure that you use the curve for S-P Diff – find the seconds on the y-axis, read over to the S-P curve, then draw a line down to the xaxis for distance. Add the values to the table above. Now you need to create the circles from each station using Figure 13.13, a map with the three stations on it. This map includes a legend in kilometers. For each station, note the distance to the epicenter. Using a drafting compass (or alternately, tie a string to a pencil, cut the string the length of the distance to the epicenter, pin it at the station, and draw a circle, with the pencil stretched out the full distance of the string), you will create the circle. First, measure the scale on the map in Figure 13.13 in centimeters, and use that to convert your distances in kilometers to centimeters (ex. the map's scale of 100 km = 2.1 cm on your ruler, so if you had a measured distance from one station of 400 km, that would equal 8.4 cm on your ruler). For this fi example, starting at the station, use the drafting compass to make a circle that is 8.4 cm in radius. Create a circle for each of the three stations, using their diff distances to the epicenter. They should overlap (or nearly overlap) in one location. The location where they overlap is the approximate epicenter of the earthquake. Once done, answer the questions below.

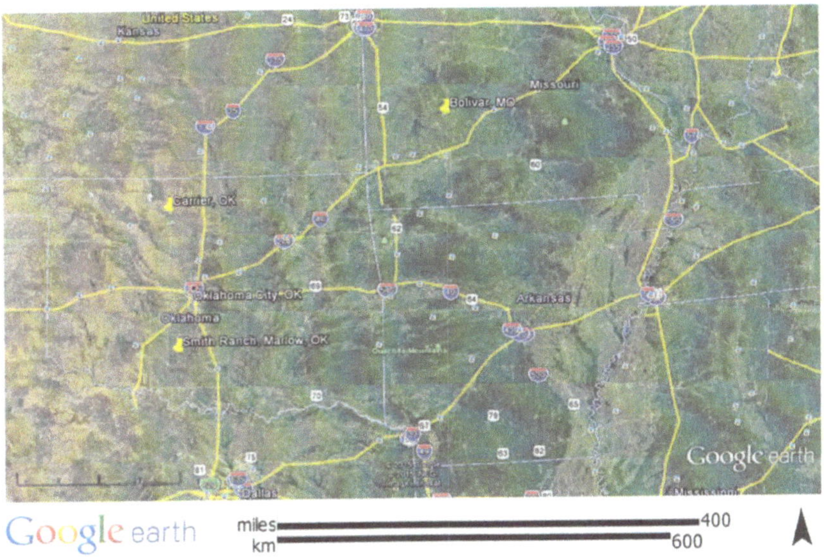

Figure 13.13

1. For Carrier, Oklahoma, what is the approximate time of the arrival of the first P-wave?
   a. 10 seconds b. 15 seconds c. 21 seconds d. 30 seconds
2. For Marlow, Oklahoma, what is the approximate time of the arrival of the first S-wave?
   a. 19 seconds b. 22 seconds c. 35 seconds d. 42 seconds
3. For Bolivar, Missouri, what is the difference between the P and S wave arrival times?
   a. 10 seconds b. 20 seconds c. 40 seconds d. 55 seconds
4. What is the approximate distance to the epicenter from Carrier, Oklahoma?
   a. 70 km b. 130 km c. 240 km d. 390 km
5. What is the approximate distance to the epicenter from Marlow, Oklahoma?
   a. 70 km b. 130 km c. 240 km d. 390 km
6. What is the approximate distance to the epicenter from Bolivar, Missouri?
   a. 70 km b. 130 km c. 240 km d. 390 km
7. Look at the location that you determined was the earthquake epicenter. Compare its location to Oklahoma City. Which direction is the epicenter located from Oklahoma City?
   a. southeast b. northwest c. northeast d. southwest

On January 12, 2010, a devastating magnitude 7.0 earthquake hit 16 miles west of Port-au-Prince, the capital of Haiti. At the following website, images are given of areas in Port-au-Prince both before the earthquake and soon after the earthquake, with a slider bar so that you can compare them. Access these images at: http://www.nytimes.com/interactive/2010/01/14/world/20100114-haiti-im- agery.html (or alternately at http://elearningexamples.com/the-destruction-in- port-au-prince-2/) and note the changes in many areas due to damage from the earthquake.

8. Examine the before and after image of the National Cathedral. Based on the changes seen within the structure, decide where this earthquake would most likely fall on the Modified Mercalli Intensity Scale. Based off this image, the most likely intensity of this earthquake would be:

   a. <IV b. V-VI c. VII d. VIII or greater

9. Residents in Port-au-Prince complained of extreme shaking during the earthquake, while residents of Santo Domingo, the capital of the Dominican Republic that sits 150 miles east of Port-au-Prince, assumed the shaking was caused by the passing of a large truck. Based on the Modified Mercalli Intensity Scale, the residents of Port-au-Prince mostly like experienced an intensity of______, while the residents of Santo Domingo experienced an intensity of___.

   a. VII, II b. VIII, III c. X, III d. X, IV

13.5 HAZARDS FROM EARTHQUAKES

Earthquakes rank among nature's most devastating occurrences, accompanied by various accompanying hazards. Seismic activity directly results in structural collapses, rendering it the most perilous hazard. The magnitude of ground shaking is influenced by various elements, such as the earthquake's size, shaking length, proximity to the epicenter, and the composition of the underlying material. Stable bedrock exhibits minimal shaking during an earthquake, making it safer than other geological substrates. Artificial fill denotes regions that have been filled for construction or garbage disposal purposes (for instance, when a hill is cleared for a shopping mall, the excavated dirt is relocated and utilized as artificial fill). Sediment remains uncompacted in regions of artificial fill; nonetheless, compaction will transpire during seismic shaking, resulting in structural failure. Artificial fill sediments exhibit behavior analogous to that of water-saturated sediments. During shaking, they may undergo liquefaction, causing the sediments to act as a fluid. Typically, friction among grains binds them cohesively. Upon the occurrence of an earthquake, water envelops each grain, hence reducing friction and resulting in liquefaction (Figure 13.14). This may pose significant threat. Seismic waves will intensify upon encountering these less resilient materials, resulting in increased damage.

Additional threats linked to earthquakes encompass fire, resulting from burst gas lines, which may be challenging to extinguish due to potential ruptures in water lines. The predominant cause of destruction during the 1906 San Francisco earthquake was fire. Earthquakes can induce tsunamis, substantial sea waves generated by the displacement of a significant volume of water during tectonic fault activity.

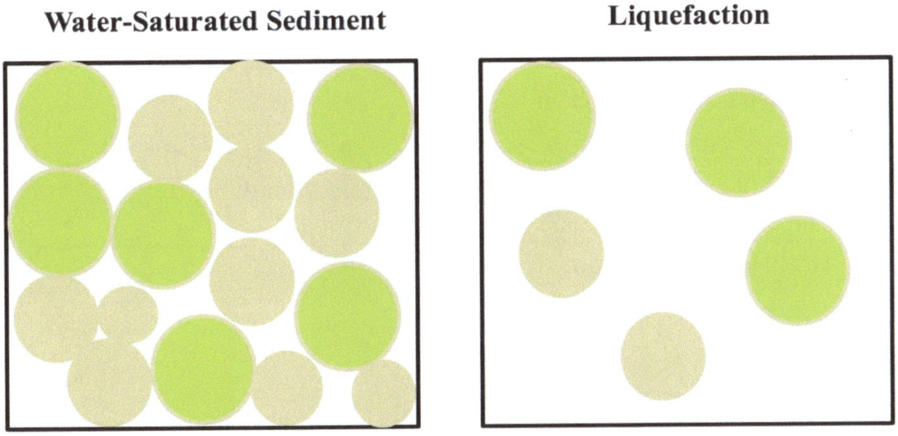

Figure 13.14 | A diagram depicting liquefaction. In the water-saturated sediment on the left, the pore (open) spaces between the grains are fi with water, but friction holds the grains together. In liquefaction, on the right, water surrounds the grains so that they no longer have contact with each other, leading them to behave as a liquid. Author: Randa Harris

The Sumatra-Andaman earthquake in 2004 triggered a tsunami in the Indian Ocean that resulted in 230,000 deaths. Earthquakes can trigger landslides in mountainous areas, and initiate secondary hazards such as fires, dam breaks, chemical spills, or even nuclear disasters like the one at Fukushima Daiichi Nuclear Power Plant in Japan. Earthquake-prone areas can take steps to minimize destruction, such as implementing strong building codes, responding to the tsunami warning system, addressing poverty and social vulnerability, retrofitting existing buildings, and limiting development in hazardous zones.

13.5.1 LAB EXERCISE

Part B Liquefaction

Download the kml file from the USGS for Google Earth found here: http://earthquake.usgs.gov/regional/nca/bayarea/kml/liquefaction.kmz (Alternately the file can be downloaded from this site: http://earthquake.usgs.gov/regional/nca/bayarea/liquefaction.php). Note that this file adds a layer of liquefaction susceptibility, with areas more likely to experience liquefaction in yellow, orange, or red. Once in Google Earth, type in San Francisco, CA. Zoom in to less than 25 miles to see the layers added and note where liquefaction is most likely, then answer the following questions. When necessary, type locations into the Search box to locate them.

10. A significant earthquake hits San Mateo, California while you are there. During the shaking you are caught indoors. Would you rather be at the US Social Security Administration Building (located at South Claremont Street, San Mateo) or with the San Mateo Park Rangers (located at J Hart Clinton Drive, San Mateo)?
    a. the US Social Security Administration Building
    b. the San Mateo Park Rangers
11. While visiting California, you become violently ill and must visit a hospital. Based off of your fears of a possible earthquake occurring, would you rather go to Highland Hospital in Oakland or Alameda Hospital in Alameda?
    a. Highland Hospital, Oakland, CA
    b. Alameda Hospital, Alameda, CA

13.6 INDUCED SEISMICITY

The number of significant earthquakes within the central and eastern United States has climbed sharply in recent years. During the thirty-six year period between 1973 and 2008, only 21 earthquakes with a magnitude of 3.0 or greater occurred. During the 5 year period of 2009-2013, 99 earthquakes of that size occurred within the same area, with 659 earthquakes in 2014 alone and well over 800 earthquakes in 2015 just in Oklahoma (see the blue and red line on the graph in Figure 13.15).

Human intervention is apparently the cause, resulting in **induced seismicity** (earthquakes caused by human activities). Humans have induced earthquakes in the past (for example, impounding reservoirs has led to earthquakes in Georgia), but this rapid increase in induced seismicity has led to much current research into the problem. Evidence points to several contributing factors, all related to types of fluid injection used by the oil industry. Hydraulic fracturing, also referred to as fracking, has been used for decades by oil and gas companies to improve well production. Fluid (usually water, though other fluids are often present) is injected at high pressure into low-permeability rocks in an effort to fracture the rock. As more fractures open up within the rock, fluid flow is enhanced and more distant fluids can be accessed, increasing the production of a well. In the past, this practice was utilized in vertical wells. With the recent advent of horizontal drilling technology, the fracking industry has really taken off. Drillers can now access thin horizontal oil and gas reservoirs over long distances, highly increasing well production in rocks that formerly were not exploited, creating a boom in US gas and oil production. While there have been many reports in the public that blame fracking for all of the increased seismicity rates, this is not entirely the case.

Fracking mainly produces very minor earthquakes (less than magnitude 3), though it has been shown to produce significant earthquakes on occasion. The majority of induced earthquakes are caused by injection of wastewater deep underground. This wastewater is the byproduct of fracking, so ultimately the industry is to blame.

As wells are developed (by fracking or other processes), large amounts of waste fluid, which may contain potentially hazardous chemicals, are created. When the fluids cannot be recycled or stored in retention ponds above ground, they are injected deep underground, theoretically deep enough to not come into contact with oil reservoirs or water supplies. These wastewater wells are quite common and are considered a safe option for wastewater disposal. By injecting this water in areas that contain faults, the stress conditions on the faults change as friction is reduced, which can result in movement along faults (resulting in earthquakes).

For our lab exercise, we will focus on the state of Oklahoma, and the increased seismicity there (Figure 13.16). The USGS has focused some research on the seismicity in Oklahoma and determined that the main seismic hazard within the state is the disposal of wastewater from the oil and gas industry by deep injection, though some smaller quakes (magnitude 0.6 to 2.9) have been shown to correlate directly to fracking. A 50% increase in earthquake rate has occurred within the state since 2013. One large earthquake of 5.7 magnitude struck in November, 2011, and has been linked to an active wastewater injection site ~200 meters away. A 4.7 magnitude earthquake struck in November 2015, too.

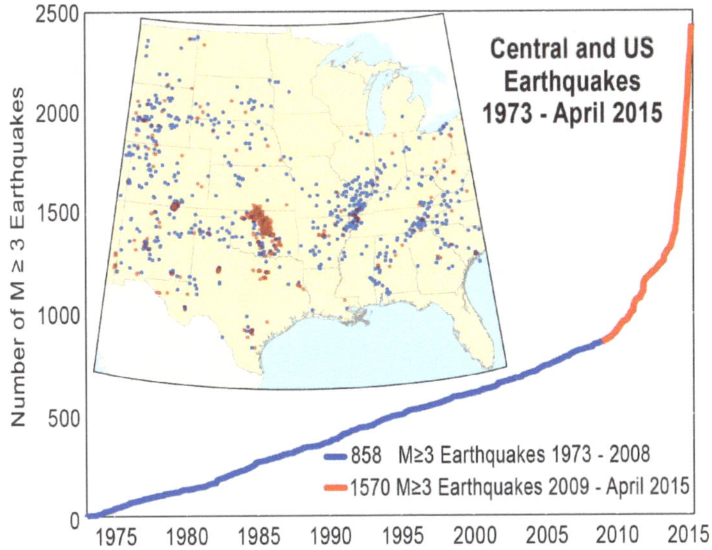

Figure 13.15 | Chart of increased seismicity of magnitude 3 or greater earthquakes within the central and eastern U.S. from 1973-2015. The spatial distribution of the earthquakes is shown on the map, with blue dots representing quakes from 1973-2008, and red dots representing quakes from 2009-2015.

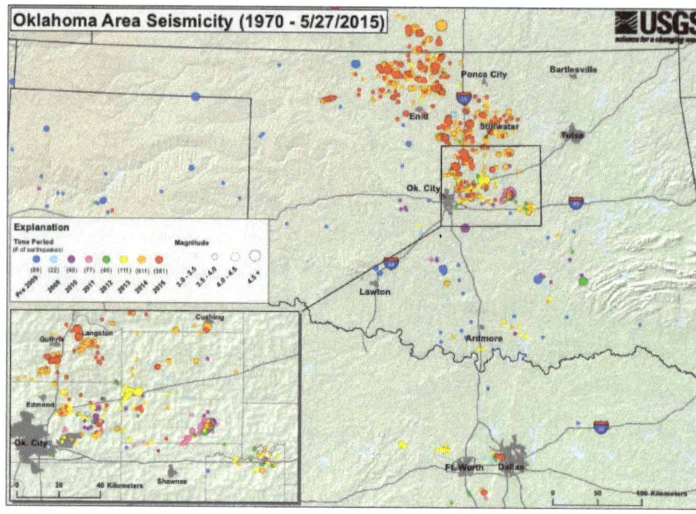

Figure 13.16 | Earthquakes that occurred within Oklahoma from 1970 – 5/27/15 are depicted above. Please note that the colors indicate year and the size indicates magnitude (see legend on image). The inset image is a close-up view of the outlined box.

13.6.1 LAB EXERCISE

Part C – Induced Seismicity

The table below contains data regarding the number of fracking wells within the state of Oklahoma and the number of significant earthquakes (magnitude 3 or greater) that have occurred since 2000. Before answering the questions for this lab exercise, plot the information in the table below on the graph that is provided; note that the graph has two y-axes, one for the number of fracking wells and the other for the number of earthquakes.

Table 13.2

Year	# of Fracking Wells in Oklahoma	# of Earthquakes greater than M 3
2000	0	0
2001	0	0
2002	0	3
2003	0	0
2004	0	2
2005	0	1
2006	0	2
2007	0	1

2008	1	2
2009	4	20
2010	1	43
2011	637	63
2012	1,568	34
2013	1,939	109
2014	3,296	585
2015	1,749	850

(From: http://www.oudaily.com/news/oklahoma-reports-surge-in-earthquakes-during/ arti- cle_79a364da-a1d4-11e5-894a-5ba84c8399c1.html)

Note: Information on number of fracked wells was obtained by SkyTruth through accessing FracFocus. Oklahomans are required to report all fracked wells, but the site was only created in 2011, so some wells may have not been retroactively added pre-2011. Seismic data was obtained through the USGS.

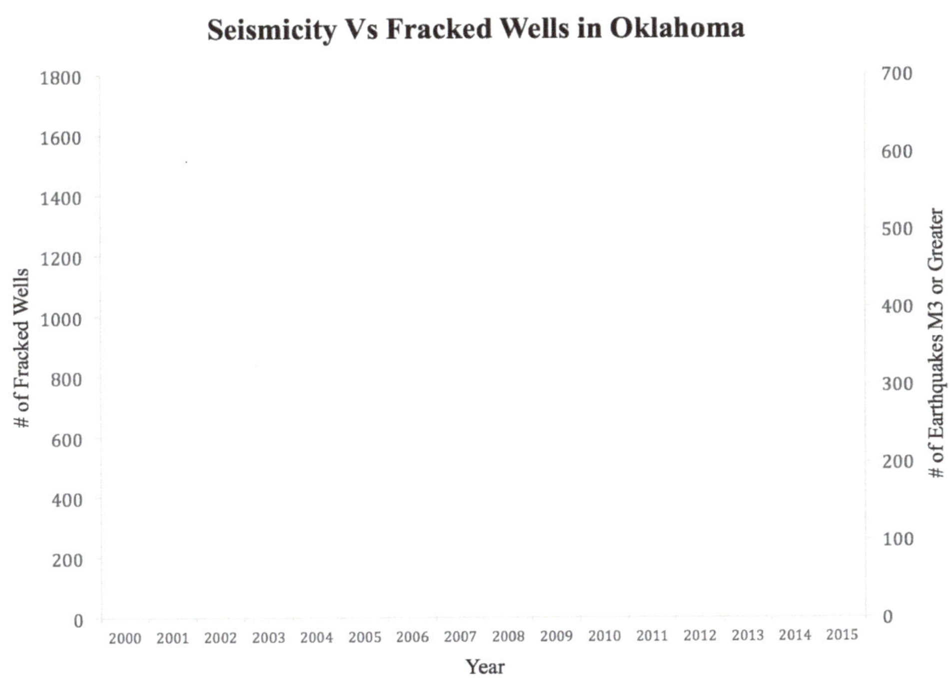

12. After what year does the number of magnitude 3 or greater earthquakes begin to rise significantly?

    a. 2007 b. 2009 c. 2011 d. 2015

13. After what year does the number of fracking wells begin to rise significantly?

    a. 2007 b. 2009 c. 2011 d. 2015

14. Based on the graph that you constructed, do significant earthquakes and the number of fracking wells appear to be related?

    a. Yes b. No

The exercises that follow use Google Earth. Let's start by examining the 1906 earthquake that hit Northern California. Access the following website: http://earthquake.usgs.gov/regional/nca/virtualtour/

There are several links of interest here. Spend some time familiarizing yourself with the site. Scroll down to the section entitled "The Northern California Earthquake, April 18, 1906" and open the link. The San Andreas Fault is ~800 miles long, located in California. In 1906, a major earthquake occurred along a portion of the fault. Scroll down and check out the Rupture Length and Slip.

15. How long was the rupture length (the length of the fault that was affected)?

    a. 25 miles b. 74 miles c. 198 miles d. 296 miles e. 408 miles

Horizontal slip, or relative movement along the fault, ranged from 2-32 feet. To envision this, imagine that you are facing an object directly across from you. Suddenly, it moves up to 32 feet to your right! Horizontal slip is shown along the rupture as a histogram. Check out all the measurements along the fault by clicking the Rupture Length and Slip on the Google Earth link.

16. Locate the epicenter of the 1906 quake. Does the amount of horizontal slip decrease faster along the northern end or the southern end of the rupture?

    a. northern end of the rupture b. southern end of the rupture

Go back to the "The Northern California Earthquake, April 18, 1906" page and scroll down to check out the Shaking Intensity. If your map begins to get difficult to read, remember that by clicking on a checked box in the Places folder, you can remove prior data. Use the search box to display the desired location.

17. What was the shaking intensity like in Sacramento?

    a. light b. strong c. severe d. violent e. extreme

18. What was the shaking intensity like in Sebastopol?

    a. light b. strong c. severe d. violent e. extreme

Navigate back to the main page and select "Earthquake Hazards of the Bay Area Today." Check out the Earthquake Probabilities for the Bay Area.

19. Based on the map, would you be more likely to experience an earthquake of magnitude >6.7 by 2031 if living in the northwest Bay Area or southeast Bay Area?

    a. northwest b. southeast

Go back to the "Earthquake Hazards of the Bay Area Today" page and check out the Liquefaction Susceptibility in San Francisco. Look at the overall trend in the areas affected by liquefaction.

20. Based on the liquefaction map, are areas more dangerous inland or along the coast?
    a. inland b. along the coast

13.7 STUDENT RESPONSES

1. For Carrier, Oklahoma, what is the approximate time of the arrival of the first P-wave?
   a. 10 seconds b. 15 seconds c. 21 seconds d. 30 seconds
2. For Marlow, Oklahoma, what is the approximate time of the arrival of the first S-wave?
   a. 19 seconds b. 22 seconds c. 35 seconds d. 42 seconds
3. For Bolivar, Missouri, what is the difference between the P and S wave arrival times?
   a. 10 seconds b. 20 seconds c. 40 seconds d. 55 seconds
4. What is the approximate distance to the epicenter from Carrier, Oklahoma?
   a. 70 km b. 130 km c. 240 km d. 390 km
5. What is the approximate distance to the epicenter from Marlow, Oklahoma?
   a. 70 km b. 130 km c. 240 km d. 390 km
6. What is the approximate distance to the epicenter from Bolivar, Missouri?
   a. 70 km b. 130 km c. 240 km d. 390 km
7. Look at the location that you determined was the earthquake epicenter. Compare its location to Oklahoma City. Which direction is the epicenter located from Oklahoma City?
   a. southeast b. northwest c. northeast d. southwest
8. Examine the before and after image of the National Cathedral. Based on the changes seen within the structure, decide where this earthquake would most likely fall on the Modified Mercalli Intensity Scale. Based off this image, the most likely intensity of this earthquake would be:
   a. <IV b. V-VI c. VII d. VIII or greater
9. Residents in Port-au-Prince complained of extreme shaking during the earthquake, while residents of Santo Domingo, the capital of the Dominican Republic that sits 150 miles east of Port-au-Prince, assumed the shaking was caused by the passing of a large truck. Based on the Modified Mercalli Intensity Scale, the residents of Port-au-Prince mostly like experienced an intensity of______, while the residents of Santo Domingo experienced an intensity of___.
   a. VII, II b. VIII, III c. X, III d. X, IV
10. A significant earthquake hits San Mateo, California while you are there. During the shaking you are caught indoors. Would you rather be at the US Social Security Administration Building (located at South Claremont Street, San Mateo) or with the San Mateo Park Rangers (located at J Hart Clinton Drive, San Mateo)?
    a. the US Social Security Administration Building
    b. the San Mateo Park Rangers

11. While visiting California, you become violently ill and must visit a hospital. Based off of your fears of a possible earthquake occurring, would you rather go to Highland Hospital in Oakland or Alameda Hospital in Alameda?

    a. Highland Hospital, Oakland, CA

    b. Alameda Hospital, Alameda, CA

12. After what year does the number of magnitude 3 or greater earthquakes begin to rise significantly?

    a. 2007 b. 2009 c. 2011 d. 2015

13. After what year does the number of fracking wells begin to rise significantly?

    a. 2007 b. 2009 c. 2011 d. 2015

14. Based on the graph that you constructed, do significant earthquakes and the number of fracking wells appear to be related?

    a. Yes b. No

15. How long was the rupture length (the length of the fault that was affected)?

    a. 25 miles b. 74 miles c. 198 miles d. 296 miles e. 408 miles

16. Locate the epicenter of the 1906 quake. Does the amount of horizontal slip decrease faster along the northern end or the southern end of the rupture?

    a. northern end of the rupture b. southern end of the rupture

17. What was the shaking intensity like in Sacramento?

    a. light b. strong c. severe d. violent e. extreme

18. What was the shaking intensity like in Sebastopol?

    a. light b. strong c. severe d. violent e. extreme

19. Based on the map, would you be more likely to experience an earthquake of magnitude >6.7 by 2031 if living in the northwest Bay Area or southeast Bay Area?

    a. northwest b. southeast

20. Based on the liquefaction map, are areas more dangerous inland or along the coast?

    a. inland b. along the coast

14 Chapter | PHYSIOGRAPHIC PROVINCES

14.1 INTRODUCTION: AN OVERVIEW

A road trip throughout the continental United States reveals notable variations in landscape, including terrain, rock formations, soil types, geological features, and vegetation, all seen from the highway. Regions exhibit diverse geologic histories, ranging from the rugged coastline of New England to the level plains of the Midwest, the jagged peaks of the Rocky Mountains, and the volcanoes of the Pacific Northwest. These observable variations across the United States can be categorized into distinct physiographic provinces. Physiographic provinces are characterized by their unique landforms, geological features, and rock assemblages.

The state of Georgia exhibits a variety of geological features, ranging from the Appalachian Mountains in the north, to the distinctive red clay of Middle Georgia, and extending to the flat Coastal Plain. In our examination of Georgia's physiographic provinces, we will also investigate the regional geological resources that benefit the state and the principal river systems that supply water to it. Ultimately, the instruments and expertise acquired from this lab manual will facilitate your reconstruction of the diverse geologies of physiographic provinces.

14.2 PHYSIOGRAPHIC PROVINCES OF THE UNITED STATES OF AMERICA

The physiographic provinces of the United States of America can be broken into three different broad areas: Western, Central, and Eastern regions (Figure 14.1). Adjacent provinces will share features or will at least be affected by the geologic events that define the nearby region. The Western Provinces are shaped by relatively young events (Post-Paleozoic), which are mostly the result of an active plate tectonic margin (the edge of the continent is also the edge of a tectonic plate).

The Eastern Provinces are a mix between young and old geologic events. Most of the individual provinces within this region result from the deformation of ancient mountain building as well as the recent passive sedimentary buildup observable today. Since the Central Provinces of the country has been largely shielded from tectonic activity, this area is flatter and less deformed than the United States' coasts. Before the individual provinces are discussed in depth, it will be helpful to review the geologic time scale that was presented in Chapter 1.

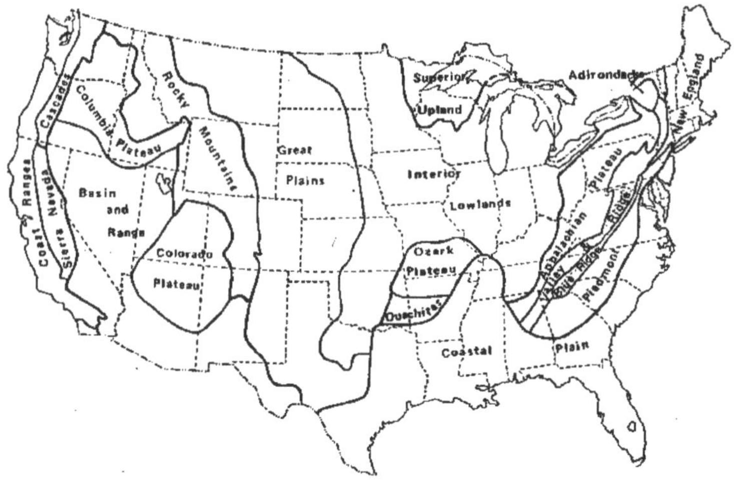

Figure 14.1 | Physiographic provinces of the continental United States of America.

14.2.1 Western U.S.

The geologic provinces in the Western United States occupy roughly a third of the country and stretch from the Rocky Mountains to the Pacific Ocean. We can group this tectonically active area into four types of provinces: the Rocky Mountains, the Coastal Pacific Mountain System, Interior Plateaus, and the extensional Basin and Range.

The **Rocky Mountains** are an immense range that stretches from New Mexico to Northern Canada. The Rocky Mountains first started to form during the Late Paleozoic, but their main growth occurred during the Laramide Orogeny during the Late Mesozoic and Early Cenozoic. An orogeny is a large-scale deformational event that is the result of the interaction between tectonic plates, in short it is mountain building. The formation of this range diff from the Continent-Continent convergent boundaries, like the Himalayas you studied in Chapter 4. The Rocky Mountains are the result of low-angle subduction, which caused deformation in the overlying plate. This striking mountain range is no longer active nor growing and has since been glaciated and eroded, which has rounded its highest peaks.

The Pacific Coast Mountains stretch from Mexico to the Arctic Circle in Alaska. Not only do the Pacific Coast Mountains diff significantly from the Rocky Mountains, but the mountain ranges within the Coastal Pacific Mountain System as also differ amongst themselves. The **Sierra Nevada Range** runs 400 miles across the middle of California and contains iconic features such Yosemite's Half Dome, which is a mecca for mountain climbers (Figure 14.2). As with the Rocky Mountains, the Sierra Nevada Range is the byproduct of convergence and subduction. During the same time as the growth of the Rocky Mountains (Late Mesozoic), the subduction led to the formation of volcanoes and the intrusion of massive granite batholiths.

Figure 14.2 | Half Dome at Yosemite National Park, which is part of the Sierra Nevada Range.

The volcanoes have long since eroded away exposing the batholiths at the surface. The Sierra Nevada topography is relatively recent (within the last 5 million years), as these rocks have been deformed by extensional stress, creating large fault blocks that build the steep face of the range. The **Cascade Range** is the northern portion of the Pacific Coast Mountain Range that spans an area from Northern California to Canada. The Cascade Range is the result of modern subduction of the Juan de Fuca and Gorda Plates under the North American Plate. Whereas the Sierra Nevada range is the eroded exposed core of a Mesozoic chain of volcanoes, the Cascade Range is modern and active. This chain encompasses thousands of volcanoes including well-known landmarks such as Mount Rainier and Mount Saint Helens.

Between the Rocky Mountains and the Pacific Coast Mountain Range is a large broad area containing a great diversity of geologic features. The Northernmost portion between the Cascades and the Rocky Mountains is the **Columbia Plateau.** The Columbia Plateau is a large igneous province dominated by Cenozoic volcanic rocks that are a product of the Yellowstone Hotspot. This hotspot produced several massive, though infrequent, eruptions; the largest eruption covering over 50,000 square miles. The Yellowstone Hotspot is still active and supplies the heat that feeds the classic geysers and hot springs throughout Yellowstone National Park (Figure 14.3). Southeast of this province is the **Colorado Plateau**, which is a broad flat expanse that considering its neighboring provinces, is remarkably un-deformed.

Figure 14.3 | Hot Spring in Yellowstone National Park within the Columbia Plateau Province.

The Colorado Plateau is mostly composed of horizontal Paleozoic and Mesozoic sedimentary rocks that contain abundant fossils, including the massive bone beds of Dinosaur National Monument. During the Laramide Orogeny this area was dramatically uplifted, allowing rivers to erode downward thus producing immense canyons, such as Grand Canyon National Park. The southeastern portion of the expanse between the Mountain Ranges is called the **Basin and Range Province.** This area can easily be recognized by the abundant Horsts and Grabens. Which you may remember are series of elevated plateaus and low basins produced from the extensional deformation of abundant paired and mirrored normal faults. There are multiple hypotheses for the source of the extension, which range from heating from the mantle to movement along the San Andres Fault.

14.2.2 Eastern U.S.

The physiographic provinces that occupy the eastern third of the country range from New England to Texas, wrapping around the Gulf of Mexico. These provinces can easily be split into three regions, such as the provinces associated with the building of the Appalachian Mountains, provinces built by much older orogenies in the Northeast, and provinces created from more recent passive build-up of sediments along the coast.

The Appalachian Mountains are the product of multiple collisions with small island chains during the Paleozoic, culminating with a collision with the supercontinent of Gondwana during the formation of Pangaea. These ancient mountains have been significantly eroded down to rolling hills that span most of the Southeastern United States as seen in Great Smoky Mountains National Park (Figure 14.4). The Appalachian Mountains can be further divided into four distinctive physiographic provinces based on their topography and geology.

Figure 14.4 | Appalachian Mountains within Great Smoky Mountains National Park, which contains multiple physiographic provinces.

The core of the Appalachian Mountains is referred to as the Valley and Ridge Province, characterized by extensively folded and faulted sedimentary rock. Fossils in these rocks suggest they are of Paleozoic era, characterized by robust Pennsylvanian-age sandstones constituting the ridges, whereas the valleys are comprised of less durable Devonian and Cambrian shales. East of the Valley and Ridge is the Blue Ridge Province, which was elevated concurrently with the Valley and Ridge but consists of somewhat older igneous and metamorphic rock. Absolute dating of these rocks reveals that the Blue Ridge was created in the Proterozoic and subsequently modified in the Middle Paleozoic. The geography of these two provinces varies, with the more homogeneous rocks of the Blue Ridge generating irregular peaks, valleys, and ridges, in contrast to the resistant and non-resistant rocks of the Valley and Ridge, which yield more consistently parallel ridges.

Figure 14.5 | Stone Mountain within the Piedmont Province.

Southeast of the Blue Ridge is the Piedmont Province, which is typified by much lower, rolling hills along with small isolated mountains such as Stone Mountain in Georgia (Figure 14.5). The Piedmont is composed of igneous and metamorphic rock from the cores of long eroded mountain chains ranging from the Proterozoic to the end of the Paleozoic period. On the adjacent west side of the Valley and Ridge province is the **Appalachian Plateau**, which is an uplifted and largely un-deformed region analogous to the Colorado Plateau. The rocks in this province are similar in age and lithology to those in the Valley and Ridge, but the preservation of fossils is enhanced because of the absence of extensive folding and faulting. In addition, the Appalachian Plateau contains abundant geological resources, including coal in West Virginia and iron in Alabama.

The physiographic provinces in the northeastern United States are very similar to the Piedmont and Blue Ridge Provinces. The **New England Province** is similar to the Piedmont in that it is composed of Late Proterozoic and Paleozoic intrusive igneous and metamorphic rocks. The major difference between these two provinces is in their history following exposure, with the New England Province showing extensive weathering and erosion from glaciers that is absent from its southeastern counterpart. The **Adirondack Mountain Province** located in upstate New York is similar to the Blue Ridge in that they are both composed of igneous and metamorphic rock. However, the Adirondacks are fairly unique considering that they are a circular rather than linear range of mountains. The rocks themselves are billion year old remnants of the building of a Proterozoic supercontinent called Rodinia. The actual mountains in the range are quite young and represent Late Cenozoic uplift, which exposed these ancient rocks.

The last eastern physiographic province is the **Coastal Plain** that spans a vast area along the Atlantic Ocean and Gulf of Mexico from New England to Texas. This province is composed of the sediment that has accumulated since the rifting of Pangaea when the eastern edge of the continent became tectonically inactive. The sediment was derived from the continent and was deposited in shallow marine sedimentary environments with abundant, mature sediment and marine fossils.

14.2.3 Central U.S.

The central third of the country is mostly flat, un-deformed, and dominated by sequences of sedimentary rocks. We separate this broad expanse into two physiographic provinces based largely on the source of the sediment. The **Interior Lowlands Province** covers the Midwestern states and consists largely of un-deformed Paleozoic marine rocks (limestones and shales) that have been since carved and shaped by the activity of glaciers and rivers. This description may not seem as geologically exciting as those to the east or west, but spectacular cave developments can occur in these marine limestones, such as those from Mammoth Cave, Kentucky (Figure 14.6).

Figure 14.6 | Mammoth Cave National Park in Central Kentucky.

To the west of the Interior Lowlands Province is the **Great Plains Province**, which is composed of sediments that eroded from the Rocky Mountains and are, therefore, substantially younger (Mesozoic and Cenozoic) with a higher proportion of clastic material. As with Interior Lowlands, the Great Plains have since been shaped by glaciers and rivers.

14.2.4 LAB EXERCISE

Materials

We will explore the various physiographic provinces of the United States of America by looking closely at the National Parks that showcase iconic geologic and topological features within each region. For each park, examine the area using Google Earth. To get a better view of the features, making sure to zoom in and out and also click on multiple photographs posted (make sure that photos are checked in the layers box). More information about these and other parks can be found at www.nps.gov.

This lab is also cumulative in that you may need to review the material presented in previous chapters to answer the following questions.

Part A – National Parks

Crater Lake National Park

Crater Lake National Park (Figure 14.7) is located in southern Oregon and was established as a National Park in 1902. The main attraction at this park is Crater Lake, which at almost 2,000 feet is one of the deepest lakes on Earth. Search for 42 56 33.15N 122 06 14.89W and zoom out to an eye altitude of 15 miles.

1. What type of volcano is Crater Lake?

   a. Composite Volcano　　b. Shield Volcano

   c. Caldera　　d. Cinder Cone

2. What is the origin of this volcano?

Figure 14.7 | Crater Lake National Park, Southern Oregon.

a. Subduction of an oceanic plate at a Convergent Boundary

b. Subduction of a continental plate at a Convergent Boundary

c. Development of a Hotspot

d. Continental Rifting

3. Crater Lake National Park is located in which of the following physiographic provinces?

a. Rocky Mountains

b. Sierra Nevada

c. Cascades

d. Columbia Plateau

e. Colorado Plateau

f. Basin and Range

Theodore Roosevelt National Park

Figure 14.8 | Theodore Roosevelt National Park, North Dakota.

Theodore Roosevelt National Park (Figure 14.8) is located in North Dakota and was established as a National Park 1978. This park was named after Theodore Roosevelt in honor of his conservation policies that led to the establishment of the National Park System. Following the death of his wife, Roosevelt spent several years in the area that would ultimately become the park, during which time he wrote extensively about the lonely beauty of the surrounding landscape. Search for 46 58 52.55N 103 32 13.91W and zoom out to an eye altitude of 30,000 feet.

4. What is the prominent drainage pattern in this park?

   a. Radial
   b. Trellis
   c. Rectangular
   d. Dendritic

5. One of the main attractions at this park is a forest of large petrified trees. These trees were preserved by being replaced with silica from ash layers within the Triassic (Mesozoic) sandstones. Based on the geologic history and features of the area, which of the following statements about these sedimentary rocks is TRUE?

   a. The clastic sand that forms these rocks was weathered and eroded from the Appalachian Mountains.
   b. The ash was produced from volcanoes associated with the Yellowstone Hotspot.
   c. The clastic sand that forms these rocks was weathered and eroded from the Rocky Mountains.
   d. The clastic sand that forms these rocks was weathered and eroded from the Sierra Nevada Mountains.

6. Theodore Roosevelt National Park is located in which of the following physiographic provinces?

   a. Rocky Mountains
   b. Colorado Plateau
   c. Interior Lowlands
   d. Great Plains
   e. Basin and Range
   f. Columbia Plateau

Acadia National Park

Figure 14.9 | Acadia National Park in Southern Maine.

Acadia National Park (Figure 14.9) is located in Southern Maine and was preserved as a National Park in 1916. This was the first park established east of the Mississippi River and helped Maine gain the nickname "Vacationland". Search for 44 21 09.94N 68 13 23.22W and zoom out to an eye altitude of 3,000 feet.

7. Based on the history of the region containing Acadia National Park and the color of the rocks, what type of igneous rock occur in this area?

   a. Gabbro b. Granite c. Rhyolite

   d. Andesite e. Diorite

8. What type of igneous body does this structure represent?

   a. Stock b. Dike c. Sill

9. Acadia National Park is located in which of the following physiographic provinces?

   a. Interior Lowlands b. Adirondacks

   c. Great Plains d. Appalachian Plateau

   e. New England f. Piedmont

Congaree National Park

Figure 14.10 | Congaree National Park in South Carolina.

Congaree National Park (Figure 14.10) is located in South Carolina and was recently established as a National Park in 2003. Unlike most National Parks, this area was preserved for reasons other than geology. Congaree contains the oldest tract of old growth hardwood forest left in the United States. This lush ecosystem contains a diverse assemblage of animals, fungi, and plants. Search for 33 47 57.63N 80 47 49.79W and zoom out to an eye altitude of 35,000 feet.

10. What is the maturity of the river in this area?

    a. Youthful b. Mature c. Old Age

11. Zoom out to an eye altitude of 500 miles to see the source of the sediment that is accumulating in this area. This sediment iṣand if lithified would be called__.

    a. Immature, Conglomerate b. Intermediate, Sandstone c. Mature, Shale

12. Congaree National Park is located in which of the following physiographic provinces?

    a. Interior Lowlands b. Blue Ridge

    c. Piedmont d. Coastal Plain e. New England

Shenandoah National Park

Figure 14.11 | Shenandoah National Park in Virginia.

Shenandoah National Park (Figure 14.11) is located in Virginia and was established as a National Park in 1935. This park, which is located close to Washington, D.C., is a favorite of hikers containing 101 miles of the Appalachian Trail, which runs from Northern Georgia to Maine. Search for 38 17 53.72N 78 40 26.42W and zoom out to an eye altitude of 25,000 feet.

13. Look over the region and examine the mountains, note their shape, ground cover, and height. Then Search for 43 48 25.03N 110 50 26.19W to examine Grand Tetons National Park. How do the Shenandoah Mountains compare to the Rocky Mountains?

    a. The Shenandoah Mountains are shorter.

    b. The Shenandoah Mountains are more rounded.

    c. The Shenandoah Mountains have more vegetation.

    d. All of the above.

14. Based on these observations, we can conclude that the mountains in Shenandoah National Park have undergone____erosion and are_______the Rocky Mountains.

    a. more, older than

    b. less, younger than

    c. the same amount of, the same age as

15. Shanandoah National Park is located in which of the following physiographic provinces?
    a. Blue Ridge
    b. Piedmont
    c. Appalachian Plateau
    d. Valley and Ridge
    e. Adirondack

Black Canyon of the Gunnison National Park

Figure 14.12 | Black Canyon of the Gunnison National Park in Colorado.

Black Canyon of the Gunnison National Park (Figure 14.12) is located in Western Colorado and was established as a National Park in 1999. This park is often overshadowed by the Grand Canyon, but is striking in its own right. The Gunnison River has a high gradient, which has produced an incredibly steep canyon. In fact, it is called Black Canyon not because of the color of the rocks, but because of the dark shadows produced by the steep walls of the canyon. Search for 38 34 43.18N 107 43 43.74W and zoom out to an eye altitude of 30,000 feet.

16. What type of weathering is primarily responsible for the formation of this canyon?
    a. Frost wedging
    b. Chemical weathering
    c. Mechanical weathering from air
    d. Mechanical weathering from water
17. We can measure the rate of erosion in this canyon at about 0.01 inches/year. Based on this rate, when did the canyon start to erode? (Hint: Measure the difference in elevation from the bottom of the canyon at the latitude and longitude given above and the top, measured at the road on the edge of the canyon due south of the previous point. Then divide by the rate of erosion. Make sure to use the correct units.)
    a. 750,000 years
    b. 1,200,000 years
    c. 2,700,000 years
    d. 3,500,000 years
18. Black Canyon of the Gunnison National Park is located in which of the following physiographic provinces?
    a. Blue Ridge
    b. Colorado Plateau
    c. Appalachian Plateau
    d. Rocky Mountains
    e. Basin and Range
    f. Columbia Plateau

14.3 GEOLOGY OF GEORGIA

Georgia serves as an exceptional natural laboratory for geological study. The rocks in this state encompass nearly a billion years of history, allowing us to examine all the subjects outlined in this lab handbook. The state encompasses mountains, coasts, geological folds, faults, seismic activity, fossils, a variety of rocks, and indications of previous volcanic eruptions. Consistent with its geological diversity, Georgia encompasses several physiographic provinces, as previously stated..

The northwestern portion of the state is within the Valley and Ridge Province and shows the characteristic sandstone ridges with folded and deformed shale within valleys. As you might expect, the shale is relatively soft (which is why they show more deformation) and erodes quickly underneath the massive sandstones. This causes the sandstones to break and tumble downhill making the Rock Cities that are a tourist attraction surrounding Lookout Mountain. The sands are Pennsylvanian in age and interbedded within them are thick coal deposits of ancient forests. These deposits were mined in the past, but for the most part are not currently active. The northeastern corner of the state is in the Blue Ridge physiographic province, which contains mountains consisting of ancient igneous and metamorphic rock. This area of Georgia contains large protected areas that preserve its natural beauty (such as Chattahoochee National Forest) and contains spectacular waterfalls as seen at Tallulah Falls State Park (Figure 14.14).

Figure 14.13 | Physiographic provinces of Georgia.

Figure 14.14 | Tallulah Falls State Park in Northern Georgia.

The central portion of the state is within the Piedmont Province that consists of rolling hills of igneous and metamorphic rocks punctuated with large batholiths. This region has several important geologic resources. First, the granite within the Piedmont has been mined for buildings, monuments, and memorial stones. The granite mining industry is one of the leading producers of granite within the United State, which is centered in Elberton, Georgia. The Piedmont Province has also produced gold, which can be found associated with quartz veins within saprolite (a soft and porous rock often formed by the weathering of granite). Starting around 1830, the discovery of gold in Dahlonega and Villa Rica, Georgia, led to the first major, though short-lived, gold rush.

The largest physiographic province within Georgia is the **Coastal Plain** in the southern half of the state. The transition from the Piedmont into the Coastal Plain is striking in that the topography dramatically changes from rolling hills to flat terrain. The boundary of the Coastal Plain is referred to as the **Fall Line**, which is a line of waterfalls along the boundary between the provinces caused by differences in the rate of erosion in the two provinces. The rocks and sediments in the Coastal Plain vary from Mid-Mesozoic in the central portion of the state to Late Cenozoic in the southern portion, tracking a fall in sea level through time. The Coastal Plain province has several useful geological resources. Foremost, its organic-rich soil and flat topography makes this region ideal for agriculture thus allowing for the abundant production of Georgia's staples such as peanuts, pecans, onions, cotton, and peaches. The mineral Kaolin, which is a product of the chemical weathering of feldspar, is also mined in this physiographic province. This white clay mineral is used in the production of many products including ceramics, toothpaste, cosmetics, and glossy paper—just to name a few.

However, one of the most important resources across Georgia is water. As with many parts of the country, Georgia has experienced steady population growth as well as droughts that have made discussions about water reservoirs and usage particularly vital to the future of the state. Georgia contains multiple river basins (Figure 14.15), many of which have been dammed to build large reservoirs, such as Lake Lanier. Study Figure 14.15, and identify the river basins that contain Georgia's major cities (such as Atlanta, Macon, and Savannah) as well as the river basin in which you live.

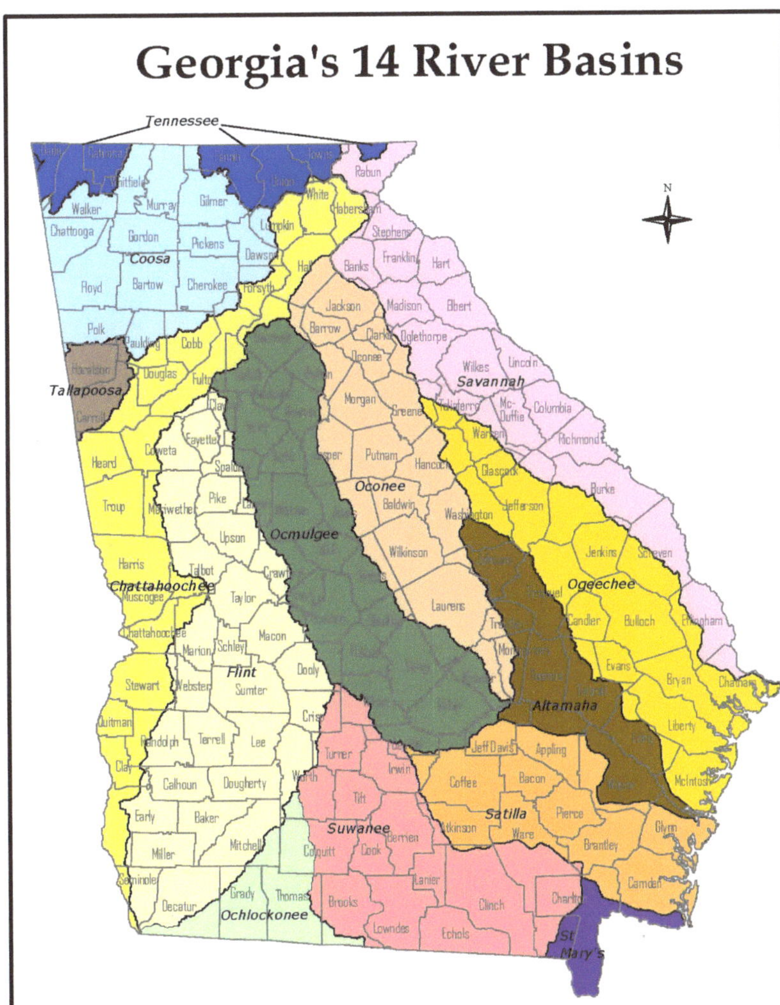

Figure 14.15 | The major river basins within Georgia.

14.3.1 LAB EXERCISE

Materials

Download the file "GA Geologic Cross Section N-S.gif" from your course website. This cross section runs north to south through the state of Georgia showing the subsurface geology across the Valley and Ridge, Piedmont, and Coastal Plain physiographic provinces. This cross section will be used to answer question 25-28. This cross-section was provided for use with permission from Geological Highway Map, Southeastern Region, American Association of Petroleum Geologists. Also, download the "Chart of Surface Time and Rock Units" legend that gives a key to the symbols and rock units included in the cross-section. We will examine the physiographic provinces of Georgia by examining a couple of Georgia's State Parks.

Part B – Georgia Provinces

Cloudland Canyon State Park

Figure 14.16 | Cloudland Canyon State Park, Dade County, Georgia.

Cloudland Canyon State Park (Figure 14.16) is located in Dade County and was established as a park in 1938. Many of the roads and civil infrastructure within Dade County were built following the designation of this area as a state park in order to allow access to the park. Cloudland Canyon State Park has multiple waterfalls as well as some of the most scenic overlooks within Georgia, especially in autumn. Search for 34 50 44.57N 85 28 40.92W in google earth and zoom out to an eye altitude of 5,000 feet.

19. Examine the walls of Cloudland Canyon, what type of rock makes up the rim of the canyon?

    a. Granite b. Shale c. Sandstone d. Gneiss

20. Zoom out to an eye altitude of 20 miles and examine the valley 6 miles to the southeast (34 43 10.37N 85 24 55.74W at the same eye altitude). What major geologic structure can be seen in this valley?
    a. Normal Fault
    b. Dome
    c. Horizontal Fold
    d. Reverse Fault
    e. Plunging Fold
21. Cloudland Canyon State Park is located in which of the following physiographic provinces?
    a. Blue Ridge
    b. Valley and Ridge
    c. Piedmont
    d. Coastal Plain

High Falls State Park

Figure 14.17 | High Falls State Park, Butts County, Georgia.

High Falls State Park (Figure 14.17) is a small park located in Butts County and was designated as a state park in 1966. This park, located between Macon and Atlanta, has the largest waterfalls in central Georgia and sits on the Fall Line. The Fall line is a line of waterfalls that is caused by differences in rock strength between two physiographic provinces. Search for 33 10 42.13N 84 01 01.11W zoom out to an eye altitude of 2,500 feet.

22. Notice that the Towaliga River Flows south, based on the position of the waterfalls, the rocks upstream are__to weathering while the rocks downstream are_____ to weathering.
    a. more resistant, less resistant
    b. less resistant, more resistant
23. Which physiographic province is located upstream of these waterfalls?
    a. Blue Ridge b. Valley and Ridge c. Piedmont d. Coastal Plain
24. Which physiographic province is located downstream of these waterfalls?
    a. Blue Ridge b. Valley and Ridge c. Piedmont d. Coastal Plain

Georgia North to South Cross-Section

Reminder: the following questions are based on the cross-section and legend provided on your courses website.

25. Examine the Cartersville Fault that is the boundary between the Valley and Ridge and Piedmont Geological Provinces. Which type of fault is the Cartersville Fault?

    a. Reverse b. Normal c. Strike-Slip

26. Which of the following statements regarding the four major faults within the Piedmont are true?

    a. The faults are all reverse like the Cartersville Fault.

    b. The faults are all normal like the Cartersville Fault.

    c. The faults are all strike-slip like the Cartersville Fault.

    d. With one exception, the faults are all similar in type to the Cartersville Fault.

    e. With two exceptions, the faults are all similar in type to the Cartersville Fault.

27. Which geologic province has abundant plutons?

    a. Valley and Ridge b. Piedmont c. Coastal Plain

28. What is the dip direction of the sedimentary layers in the Coastal Plain Province in Peach County?

    a. North b. South c. East d. West

Part C – Georgia's Natural Resources

Search for 33 18 32.95N 82 26 08.55W and zoom to an eye altitude of 8,000 feet.

29. What natural resource is being mined in this area?

    a. Gold b. Granite c. Kaolin d. Coal

30. This mine is located in which of the following physiographic province?

    a. Blue Ridge b. Valley and Ridge c. Piedmontd. Coastal Plain

31. The mineral that was weathered to produce these large deposits likely originated from which of the following physiographic province?

    a. Blue Ridge b. Valley and Ridge

    c. Piedmont d. Coastal Plain

Search for 33 27 11.53N 84 56 12.56W and zoom out to an eye altitude of 5,000 feet. To answer the following questions also refer to Figure 14.15.

32. Rainfall in this forest would be part of which of the following Georgia River Basins?

    a. Flint b. Tallapoosa c. Coosa d. Ocmulgee

    e. Oconee f. Chattahoochee g. Suwanee

33. Zoom out to at least an eye altitude of 15 miles and examine the major river in this drainage basin. What overall direction does this river flow?
    a. Northeast to Southwest b. Southwest to Northeast
    c. Northwest to Southeast d. Southeast to Northwest

34. Follow this river downstream, into what body of water does this basin ultimately flow?
    a. Atlantic Ocean b. Lake Lanier
    c. Lake Martin d. Gulf of Mexico

35. How many times has this river been dammed to produce reservoirs that are most likely used for drinking water, power, or recreation?
    a. Once b. Twice
    c. Three times d. Greater than three times

14.4 STUDENT RESPONSES

1. What type of volcano is Crater Lake?
   a. Composite Volcano b. Shield Volcano
   c. Caldera d. Cinder Cone

2. What is the origin of this volcano?
   a. Subduction of an oceanic plate at a Convergent Boundary
   b. Subduction of a continental plate at a Convergent Boundary
   c. Development of a Hotspot
   d. Continental Rifting

3. Crater Lake National Park is located in which of the following physiographic provinces?
   a. Rocky Mountains b. Sierra Nevada
   c. Cascades d. Columbia Plateau
   e. Colorado Plateau f. Basin and Range

4. What is the prominent drainage pattern in this park?
   a. Radial b. Trellis c. Rectangular d. Dendritic

5. One of the main attractions at this park is a forest of large petrified trees. These trees were preserved by being replaced with silica from ash layers within the Triassic (Mesozoic) sandstones. Based on the geologic history and features of the area, which of the following statements about these sedimentary rocks is TRUE?
   a. The clastic sand that forms these rocks was weathered and eroded from the Appalachian Mountains.
   b. The ash was produced from volcanoes associated with the Yellowstone Hotspot.
   c. The clastic sand that forms these rocks was weathered and eroded from the Rocky Mountains.

d. The clastic sand that forms these rocks was weathered and eroded from the Sierra Nevada Mountains.

6. Theodore Roosevelt National Park is located in which of the following physiographic provinces?

   a. Rocky Mountains b. Colorado Plateau c. Interior Lowlands

   d. Great Plains e. Basin and Range f. Columbia Plateau

7. Based on history of the region containing Acadia National Park and the color of the rocks, what type of igneous rock occur in this area?

   a. Gabbro b. Granite c. Rhyolite d. Andesite e. Diorite

8. What type of igneous body does this structure represent?

   a. Stock b. Dike c. Sill

9. Acadia National Park is located in which of the following physiographic provinces?

   a. Interior Lowlands b. Adirondacks

   c. Great Plains d. Appalachian Plateau

   e. New England f. Piedmont

10. What is the maturity of the river in this area?

    a. Youthful b. Mature c. Old Age

11. Zoom out to an eye altitude of 500 miles to see the source of the sediment that is accumulating in this area. This sediment is_____and if lithified would be called_____.

    a. Immature, Conglomerate b. Intermediate, Sandstone

    c. Mature, Shale

12. Congaree National Park is located in which of the following physiographic provinces?

    a. Interior Lowlands b. Blue Ridge

    c. Piedmont d. Coastal Plain

    e. New England

13. Look over the region and examine the mountains, note their shape, ground cover, and height. How do these mountains compare to the Rocky Mountains (search for 43 48 25.03N 110 50 26.19W to examine Grand Tetons National Park)?

    a. The Shenandoah mountains are shorter.

    b. The Shenandoah mountains are more rounded.

    c. The Shenandoah mountains have more vegetation.

    d. All of the above.

14. Based on these observations, we can conclude that the mountains in Shenandoah National Park have undergone____erosion and are _______the Rocky Mountains.

    a. more, older than

    b. less, younger than

    c. the same amount, the same age as

15. Shenandoah National Park is located in which of the following physiographic provinces?

    a. Blue Ridge b. Piedmont

    c. Appalachian Plateau d. Valley and Ridge e. Adirondack

16. What type of weathering is primarily responsible for the formation of this canyon?

    a. Frost wedging b. Chemical weathering

    c. Mechanical weathering from air d. Mechanical weathering from water

17. We can measure the rate of erosion in this canyon at about 0.01 inches/year. Based on this rate, when did the canyon start to erode? (Hint: Measure the difference in elevation from the bottom of the canyon at the latitude and longitude given above and the top, measured at the road on the edge of the canyon due south of the previous point. Then divide by the rate of erosion. Make sure to use the correct units.)

    a. 750,000 years b. 1,200,000 years

    c. 2,700,000 years d. 3,500,000 years

18. Black Canyon of the Gunnison National Park is located in which of the following physiographic provinces?

    a. Blue Ridge b. Colorado Plateau

    c.Appalachian Plateau d. Rocky Mountains

    e. Basin and Range f. Columbia Plateau

19. Examine the walls of Cloudland Canyon, what type of rock makes up the rim of the canyon?

    a. Granite b. Shale c. Sandstone d. Gneiss

20. Zoom out to an eye altitude of 20 miles and examine the valley 6 miles to the southeast (34 43 10.37N 85 24 55.74W at an eye altitude of 20 miles). What major geologic structure can be seen in this valley?

    a. Normal Fault b. Dome

    c. Horizontal Fold d. Reverse Fault

    e. Plunging Fold

21. Cloudland Canyon State Park is located in which of the following physiographic provinces?

    a. Blue Ridge b. Valley and Ridge

    c. Piedmont d. Coastal Plain

22. Notice that the Towaliga River Flows south, based on the position of the waterfalls, the rocks upstream are_to weathering while the rocks downstream are___to weathering.

 a. more resistant, less resistant b. less resistant, more resistant

23. Which physiographic province is located upstream of these waterfalls?

 a. Blue Ridge b. Valley and Ridge

 c. Piedmont d. Coastal Plain

24. Which physiographic province is located downstream of these waterfalls?

 a. Blue Ridge b. Valley and Ridge

 c. Piedmont d. Coastal Plain

25. Examine the Cartersville Fault that is the boundary between the Valley and Ridge and Piedmont Geological Provinces. Which type of fault is the Cartersville Fault?

 a. Reverse b. Normal c. Strike-Slip

26. Which of the following statements regarding the four major faults within the Piedmont are true?

 a. The faults are all reverse like the Cartersville Fault

 b. The faults are all normal like the Cartersville Fault

 c. The faults are all strike slip like the Cartersville Fault

 d. With one exception, the faults are all similar in type to the Cartersville Fault

 e. With two exceptions, the faults are all similar in type to the Cartersville Fault

27. Which geologic province has abundant plutons?

 a. Valley and Ridge b. Piedmont c. Coastal Plain

28. What is the dip direction of the sedimentary layers in the Coastal Plain Province in Peach County?

 a. North b. South c. East d. West

29. What natural resource is being mined in this area?

 a. Gold b. Granite c. Kaolin d. Coal

30. This mine is located in which of the following physiographic province?

 a. Blue Ridge b. Valley and Ridge c. Piedmont d. Coastal Plain

31. The mineral that was weathered to produce these large deposits likely originated from which of the following physiographic province?

 a. Blue Ridge b. Valley and Ridge c. Piedmont d. Coastal Plain

32. Rainfall in this forest would be part of which of the following Georgia River Basins?

 a. Flint b. Tallapoosa c. Coosa d. Ocmulgee

 e. Oconee f. Chattahoochee g. Suwanee

33. Zoom out to at least an eye altitude of 15 miles and examine the major river in this drainage basin. What overall direction does this river flow?
    a. Northeast to Southwest
    b. Southwest to Northeast
    c. Northwest to Southeast
    d. Southeast to Northwest
34. Follow this river downstream, into what body of water does this basin ultimately flow?
    a. Atlantic Ocean
    b. Lake Lanier
    c. Lake Martin
    d. Gulf of Mexico
35. How many times has this river been dammed to produce reservoirs that are most likely used for drinking water, power, or recreation?
    a. Once
    b. Twice
    c. Three times
    d. Greater than three times

INDEX

www.ingramcontent.com/pod-product-compliance
Lightning Source LLC
LaVergne TN
LVHW071245210926
848486LV00008B/354
9781787156418